THE ELEVENTH PLAGUE

THE
ELEVENTH
PLAGUE

The Politics of
Biological and Chemical Warfare

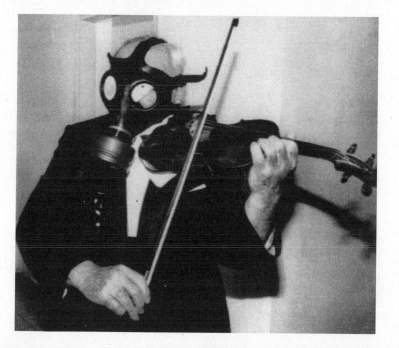

LEONARD A. COLE

A W. H. Freeman/Owl Book
Henry Holt and Company   New York

Henry Holt and Company, LLC
Publishers since 1866
115 West 18th Street
New York, New York 10011

Henry Holt® is a registered trademark
of Henry Holt and Company, LLC.

Library of Congress Cataloging-in-Publication Data

Cole, Leonard A., 1933–
      The eleventh plague : the politics of biological and chemical warfare /
Leonard A. Cole.
            p.   cm.
      Includes index.
      ISBN 0-7167-2950-4
      ISBN 0-8050-7214-4 (pbk.)
      1. Biological warfare. 2. Chemical warfare. 3. Biological weapons—Iraq.
4. Disarmament. I. Title.

UG447.8.C6523   1996   96-24094
327.1'74—dc20   CIP

Henry Holt books are available for special
promotions and premiums. For details contact:
Director, Special Markets.

First Owl Books Edition 2002

A W. H. Freeman / Owl Book

Designed by Victoria Tomaselli

Printed in the United States of America

1   3   5   7   9   10   8   6   4   2

CONTENTS

THE ELEVENTH PLAGUE

Introduction:
Testing Ourselves

"These are the ten plagues which the Holy One brought upon the Egyptians in Egypt: blood, frogs, vermin, flies, murrain, boils, hail, locusts, darkness, and the slaying of the first-born." Recited every year as part of the Passover celebration, the passage relates to the biblical story of Exodus. The plagues were God's punishment because Pharaoh refused to release the Jews from slavery.

None of the plagues was biological and chemical warfare. Did God not know about the eleventh plague? Unlikely, because God presumably is omniscient. Perhaps poison weapons seemed so vile that God wished people might never find out about them.[1] If so, the hope has been disappointed. The loss of innocence began on a sunny day in 1915.

On the afternoon of April 22, French soldiers watched from their trenches near Ypres in Belgium as a strange yellow-green cloud rolled toward them. German troops a few hundred yards away had released chlorine gas from 6000 cylinders to catch the wind blowing toward the French lines. The gas hung close to the ground and within minutes thousands of soldiers were enveloped. Gasping for breath, they began to froth at the mouth. One account described them as writhing in "agony unspeakable, their faces plum colored, while they coughed blood from their tortured lungs."[2]

This first large-scale use of a chemical in warfare presaged later attacks by all the major powers in the conflict. As the war continued, additional agents were developed, culminating with the German introduction of mustard gas in 1917. Unlike other chemical weapons, mustard affected not only the lungs, but the body surface. Even while wearing respirators, exposed troops were blinded and their skin became painfully blistered. Some died; others suffered lifelong injury. By the end of World War I, chemical agents had caused 1.3 million casualties including nearly 100,000 deaths.[3]

Revulsion about this experience prompted efforts to ban the use of chemical and biological weapons. But in the very recent past, the world was reminded that poison weapons have become more horrifying than ever.

Today's Worries

"It hurt to breathe. I could feel it in my nostrils," recalled Akio Masahata soon after he sat aboard a Tokyo subway train on March 20, 1995. "People were starting to collapse around me." He was one of 5500 passengers injured after terrorists from a religious cult released the nerve agent sarin into the subways.[4]

A tiny drop of sarin, which was developed in Germany in the 1930s, can kill within minutes after skin contact or inhalation of its vapor. That only 12 people died from the Tokyo attack was attributed to an impure mixture of agent. But agony was rife nonetheless. A news account reported that "passengers were rushed out on stretchers and lay on the ground with bubbles coming from their mouths. In some cases blood poured from their noses."[5]

The cult responsible for the sarin attack, Aum Shinrikyo, was developing biological agents as well. If a chemical attack is frightening, a biological weapon poses a worse nightmare. Bacteria, viruses, and other live agents may be contagious and reproduce. If they become established in the environment, they may increase in number. Unlike any other weapon, therefore, they can become more dangerous with the passage of time.

Certain biological agents incapacitate, whereas others kill. The dengue fever virus, for example, causes rash, headache, and body pain, but recovery after a few weeks is common. On the other hand, the Ebola virus kills as many as 90 percent of its victims. In little more than a week, it takes the body apart piece by piece.

Within days of infection, Ebola patients suffer from soaring temperature and excruciating joint and muscle pain. The throat is so sore that swallowing anything, including one's own saliva, is intolerable. The connective tissue liquefies. The skin becomes like soft bread—it can be spread apart with one's fingers and blood oozes out. Victims choke as the sloughed-off surfaces of their tongues and throats slide into their windpipes. Every body orifice bleeds. Even the eyeballs fill with blood that leaks down the cheeks.

In the final stages, victims become convulsive, splashing blood all around as they twitch, shake, and thrash to their deaths. Ebola, according to science writer Richard Preston, "is a perfect parasite because it transforms virtually every part of the body into a digested slime of virus particles."[6]

For Ebola there is no cure, no treatment. Even the manner in which it spreads is unclear: Is it only by close contact with a victim and his blood or just by breathing the air nearby? Recent outbreaks in Zaire prompted the quarantine of sections of the country until the disease ran its course. When Western doctors visited patients there, they wore respirators, protective outerwear, and gloves.

The horror is only magnified by the thought that individuals and nations would consider attacking others with such viruses. A 1995 investigation by staff members of a U.S. Senate committee indicated that Aum Shinrikyo was seeking to develop the Ebola virus as a weapon. Aum representatives went to Zaire in 1992 ostensibly to help treat Ebola victims. But the real intention apparently was to obtain virus samples, culture them, and use them in biological attacks.[7]

Interest in acquiring killer organisms for unseemly purposes is not limited to groups outside the United States. On May 5, 1995, barely six weeks after the Tokyo subway incident, Larry

Harris, a laboratory technician in Ohio, ordered bacteria that cause bubonic plague from a Maryland biomedical supply company. The company mailed him three vials of the germs.

Harris drew suspicion only when he called the firm four days after placing his order to find out why it had not arrived. Company officials wondered about his impatience and his apparent unfamiliarity with laboratory techniques, and contacted federal authorities. He was later found to be a member of a white supremacist organization. A search warrant indicated that he claimed he wanted to conduct research to counteract Iraqi supergerm-carrying rats. In November he pled guilty in federal court to mail fraud.[8] But to get the plague bacteria in the first place, he needed little more than a MasterCard.

What would Harris have done with the bacteria? If he cared to grow a biological arsenal, the task is frighteningly simple. By dividing every 20 minutes, a single bacterium gives rise to more than a billion in 10 hours. A small vial of microorganisms can yield a huge number before the end of a week. For some diseases, like anthrax, inhaling a few thousand bacteria—smaller in area than the period at the end of this sentence—is enough to kill.

Kathleen Bailey, a former assistant director of the U.S. Arms Control and Disarmament Agency, has visited several biotechnology and pharmaceutical firms. She is "absolutely convinced" that a major biological arsenal could be built in a room 15-by-15 feet, with $10,000 worth of equipment.[9]

But worry about terrorism from individuals is only one side of the threat. The other comes from governments. These inexpensive weapons appeal especially to countries with limited resources. They are aptly called the poor man's atomic bomb.

As the twentieth century ends, an unpleasant paradox has emerged. More countries than ever are signing international agreements to eliminate chemical and biological arms. Yet more are also developing these weapons despite the treaties. In 1980, about a dozen countries possessed chemical weapons. Only one country, the Soviet Union, had been named by the United States as violating the Biological Weapons Convention, which prohibits possession of biological weapons.[10]

Since then, the numbers have ballooned. "More than 25 countries are now suspected of having chemical weapons or the ability to produce them," observed John D. Holum, director of the U.S. Arms Control and Disarmament Agency, in 1994.[11] By 1995, 17 countries had been named as biological weapons suspects.[12] Five of the countries with suspected biological arms programs—Iran, Iraq, Libya, Syria, North Korea—have histories of militant behavior. Moreover, four of these five, North Korea excepted, have signed the Biological Weapons Convention. (Although a signatory, Syria has not ratified.) A 1994 Pentagon report to Congress cited instability in Eastern Europe, the Middle East, and Southwest Asia as likely to encourage even more nations to develop biological and chemical warfare capabilities.[13]

Thus, although sarin in Tokyo created headlines, the behavior of nations deserves no less attention. United States activities in the name of defense; Iraq's use of chemical weapons during the 1980s against Iran; the Iraqi chemical and biological threat to coalition forces during the 1991 Persian Gulf War; efforts to rid the world of these weapons—all are reviewed here and all underscore the extraordinary danger these weapons present.

The emphasis in this book is on biological warfare, which is deemed especially repugnant and potentially can inflict more widespread damage than chemicals. But insofar as biological and chemical issues commonly overlap, chemicals are discussed as well. The fundamental argument is that striving to eliminate both forms of weaponry is feasible and desirable. To fail in the effort is to vastly increase the likelihood of a man-made plague—from Ebola or some other gruesome agent.

Grim Realities

Dedication to biological disarmament should be enhanced by another grim truth: For many scenarios, a large population cannot be protected against a biological attack. Vaccines can prevent some diseases but, unless the causative agent is known in advance, a vaccine may be worthless. Antibiotics are effective against specific bacteria or classes of biological agents, but not against all. Moreover, the incidence of infectious disease throughout the world has been rising because of newly resistant strains of bacteria that defy treatment. In this era of biotechnology, especially, novel organisms can be engineered against which vaccines or antibiotics are useless.

An enemy is unlikely in any case to reveal in advance which bacteria or viruses it will use in an attack. Thus, besides the logistical problem of timely inoculations for millions of people, medical defenses are likely to be ineffective. Nor do physical barriers offer great comfort.

"Once a hazardous cloud has been detected," said Graham Pearson, recently head of Britain's Chemical and Biological Defense Establishment, "it is necessary to don appropriate physical protection consisting of a respirator and, depending on the information available, a protective suit or to enter collective facilities supplied with filtered air." Most biological agents have no effect against intact skin, but they would still require wearing "respiratory protection until the hazard has decreased to a safe level."[14]

Pearson presumed that the biological danger would be short-lived, expecting that sunlight and ambient temperatures would destroy the agents. But certain microorganisms can persist in an environment indefinitely. Gruinard Island off the coast of Scotland remained infected with anthrax spores for 40 years after biological warfare tests there in the 1940s.[15] Pearson's predecessor, Rex Watson, said in 1981 that if Berlin had been attacked with anthrax bacteria in World War II, the city would still have been contaminated.[16] Expecting anyone to wear protective masks for months or years is, of course, ridiculous.

Despite these realities, calls for more funding and larger biological defense programs have increased. A lesson of the Persian Gulf War, according to one analyst, is that "the United States cannot any longer neglect the minimal tasks of bw preparedness, that is, fielding adequate detectors and protective gear and preparing vaccines for known bw agents."[17] Although focused on battlefield situations, these remarks imply as well that detectors, special gear, and vaccines can protect a larger public. But the same truths hold for both the public and the military. Unless an attack organism is known in advance and is vulnerable to medical interventions, defense can be illusory.

In a broad perspective, the 1991 Persian Gulf experience may have been misleading. Iraq's biological weapons were understood to be anthrax bacilli and botulinum toxin. Both are susceptible to existing vaccines and treatments, and protection of military forces therefore seemed possible. In this case, defense preparations may have benefited from earlier military research. But the improbabilities of protection from less traditional agents deserve full appreciation. Expecting that research can come up with defenses against attack organisms whose nature is not known in advance seems fanciful. The notion is akin to believing that a large population can be defended against a nuclear attack.

Cheap, Easy, but Rarely Used

Although chemical weapons are inanimate and biological weapons are living organisms, they share several characteristics. (Toxins, which are inanimate products of living organisms, are treated as biological agents in the Biological Weapons Convention.) When biological or chemical weapons are released in the air, their effectiveness depends on meteorological conditions. Both therefore are seen as less controllable and more nasty than conventional weapons. Moreover, the Pentagon continues to link the two in its budgetary lines and command structure, the U.S. Army Chemical and Biological Defense Agency.

But biological weapons are distinctive in several ways. An army general who was involved with both systems noted that "chemical agents will cover only tens of square miles, but biological agents can blanket hundreds of thousands of square miles."[18] A recent study by the Office of Technology Assessment estimated that 1000 kilograms of sarin nerve gas released from an airplane could kill 8000 people. But a release of only 100 kilograms (about 220 pounds) of anthrax bacteria could result in 3 million deaths.[19]

Unlike most chemicals, biological agents cannot be separated from a natural habitat, and they may not be recognized until having caused widespread infection. The likelihood of early warning and detection is remote. The problem is accentuated by the fact that novel organisms can be created for which there are no vaccines or treatments.

Biological agents are also cheaper than other weapons. A United Nations panel received information in 1969 that "for a large-scale operation against a civilian population, casualties might cost about $2000 per square kilometer with conventional weapons, $800 with nuclear weapons, $600 with nerve-gas weapons, and $1 with biological weapons."[20]

Yet for all the ease and low cost, poison weapons have rarely been used. The reason in part lies with the sense of repugnance they generate, which was underscored by the World War I experience and the 1925 Geneva Protocol that followed.

In prohibiting the use of chemical and bacteriological weapons in war, the protocol proclaims that such weapons have been "justly condemned by the general opinion of the civilized world."[21] The language fostered the notion that chemical and biological weapons were morally unacceptable. Although the protocol contains no provisions for enforcement, its moral authority helped deter the use of chemical or biological agents in subsequent decades. Japan is the only country known to have used biological weapons, in attacks on China in the late 1930s and the 1940s. Japanese aircraft dropped bacteria-filled bombs, feathers, and cotton wadding, victimizing thousands of Chinese, 700 from plague alone.[22]

Chemical infractions have been more frequent. In the 1930s, Italy used chemical weapons against Ethiopia, as did Japan against China and Egypt against Yemen in the 1960s. Still, the international norm against their use largely prevailed. Not until the Iran-Iraq war in the 1980s did a large-scale extended violation of the Geneva Protocol take place.

A Weakened Ethos of Repugnance

The Persian Gulf crisis highlighted a process that began during the Iran-Iraq war—erosion of the traditional repugnance toward biological and chemical weapons. Not since World War I had the possibility of a large-scale gas attack against Western forces seemed so imminent to the public.[23] But by 1991, pictures of American and other coalition troops wearing gas masks and protective gear in preparation for battle signaled the change of mood. Media inquiry about Iraq's unethical and illegal activities was sparse. Abhorrence of gas and germ weapons had yielded to unhappy resignation.

The shift of ethical focus was a consequence of policies by the United States and other powers begun years earlier. It derived from perceived international interests and new opportunities born of biochemical technology. How the United States arrived at that point is best understood by a reminder of events since World War II.

Although the United States did not ratify the Geneva Protocol until 1975, it had long been a signatory and in effect abided by its terms. (Treaty ratification requires concurrence of two-thirds of the Senate present.) The treaty prohibited the use in war of chemical and bacteriological weapons, but not their possession. Accordingly, the United States maintained substantial arsenals during the 1950s and 1960s. Indeed, fear of the Soviet Union prompted a full-scale program of weapons development—chemical, biological, nuclear, and conventional.[24]

In 1969, President Richard Nixon reordered the assumption that weapons-in-kind were needed for deterrence. He terminated the U.S. biological arsenal, reasoning that the country could respond to a biological threat with nonbiological weapons.[25]

In the next decade, the Biological Weapons Convention (BWC) was established, and the United States ratified the Geneva Protocol. But by the early 1980s, fears that the Soviets were violating the BWC, coupled with new technological possibilities, prompted a U.S. reassessment. A 1985 presidentially appointed Chemical Warfare Review Commission, for example, observed that "rapid advances of genetic technology . . . offer the predictable likelihood of new agents being developed." It recommended "an expanded research and development effort on both chemical weapons and defensive measures, and greater research into defenses against toxins and biological agents that the Soviets are developing."[26]

At the same time, the United States was engaging in an exercise of political utility at the expense of a competing ethical principle. Like most of the international community, it offered no substantive protests against Iraq's use of chemical weapons in the Iran-Iraq war. The United States tilted toward Iraq in fear of Khomeini-inspired Islamic fundamentalism. By the time Iran agreed to a cease-fire in 1988, Iraq had accomplished something no other nation in history had. Not only did it use chemical weapons for more than four years with impunity, it created a perception that these weapons helped determine the outcome of a war.

While this perception was taking hold among military planners throughout the world, U.S. policy makers continued to avoid meaningful criticism of Iraq. They seemed more interested in exploring the possibility of new weapons than in enforcing old treaties. The annual budget for chemical warfare and biological defense activities climbed from $160 million in 1980 to more than $1 billion in 1986.

The expanding U.S. program itself prompted legal and safety questions. At the end of the decade, a congressional inquiry found a "disturbing record of safety problems" at military chemical and biological research facilities.[27] Reports of accidents had been concealed, and batches of dangerous microorganisms were inexplicably missing from laboratories.

Criticisms of the program were offered against a backdrop of new information about earlier military experiments. The army had conducted biological warfare tests during the 1950s and 1960s that included spraying much of the country with bacteria and chemical materials. The tests were rehearsals for an attack by or against an enemy with more lethal agents. But they presented risks to the health of millions of citizens who unwittingly were exposed to the army's agents.

Current U.S. biological and chemical activities, problematic in their own right, have thus been shadowed by the earlier tests. The history of the U.S. military programs suggests that the more expansive they are, the more the public has been endangered. Ironically, then, the public has reason to worry not only about biological and chemical weapons in other parts of the world, but about programs at home to develop defenses against them.

Prohibition versus Proliferation

Impetus to do away with these weapons had previously received a boost with the establishment of the 1972 Biological Weapons

Convention. Unlike any earlier international agreement, the BWC sought total elimination of a major weapons system. By mid-1996, 137 states were parties to the convention. They had agreed to "prohibit and prevent the development, production, stockpiling, acquisition or retention" of biological or toxin weapons, and the means to deliver them for hostile purposes.[28] Article I does permit activities for "prophylactic, protective or other peaceful purposes," which is understood to mean that defensive work is allowable. But the treaty is explicit in its prohibition of offensive capabilities.

The BWC mentions as well the hope that its establishment will lead to an agreement to eliminate chemical weapons. Two decades later the hope was realized. The Chemical Weapons Convention (CWC) was opened for signature in January 1993.[29] By the end of 1996, 160 states had signed. In November, the 65th country deposited instruments of ratification, which meant the treaty would enter into force in April 1997. The U.S. Senate had not yet ratified, however. (Were the United States not to become a party, the effectiveness of the CWC would be gravely weakened.)

The two treaties contain important differences that are explored in later chapters. But they share the overarching presumption that total biological and chemical disarmament is possible. A large presumption, it has been endorsed by a great majority of the world's nations. The driving wisdom behind the treaties is that only when biological and chemical weapons are unavailable can we feel secure about their not being used.

The tension prompted by the divergent tracks of prohibition and proliferation is mirrored by competing advocacies in the United States. Several scholars, government officials, and members of the attentive public vigorously support international agreements to prevent a biological or chemical war. They urge strong treaties and reduced defense budgets. In *Preventing a Biological Arms Race*, historian Susan Wright, for example, argues that the Biological Weapons Convention should be reinforced and that governments should "abjure research likely to produce results relevant to weapons applications."[30]

Others believe that the treaties are virtually irrelevant and that the United States needs to confront an adversary's gas and germ weapons with its own. Thus, Joseph D. Douglass and Neil C. Livingstone, in *America the Vulnerable,* describe the Biological Weapons Convention as a "sham." They believe the United States should withdraw because "the best deterrent to the use of C/B weapons is a similar capability in the hands of the other side."[31]

Even many who do not endorse an offensive capability oppose Wright's call for lower military expenditures. "The end of the Cold War," said Congressman Jim Hansen of Utah in a plea for more defense research, "has intensified, not diminished, the threat of biological warfare."[32] Whatever their advocacies, however, all observers recognize the growing importance of the issue.

Interest in biological weapons is manifested in the publication during the past decade of some 16 books on the subject.[33] The commander of the army's Chemical and Biological Defense Agency noted that "the biological threat has been recently singled out as the one major threat that still poses the ability for catastrophic effects on a theater-deployed force."[34]

The immediacy of the issue was highlighted in March 1996 when a NATO workshop considered the formidable challenges posed by biological weapons and biological terrorism. At the same time, Senator Sam Nunn chaired a Senate committee hearing on what he called "the new era" since the Tokyo nerve-gas attack. "Zealots and terrorists are now increasingly willing to do the unthinkable."[35]

Fostering the Ethos of Repugnance

Judging from the tenor of descriptions, few human enterprises are more despicable than biological warfare. Historian Barton Bernstein observes that the idea of deliberately spreading disease evokes "widespread revulsion." For microbiologist Robert Sinsheimer, biological weapons "are simply unfit for any human use at any time in any cause."[36] A British biologist called developing biological weapons, let alone using them, "a crime against humanity."[37] Richard Nixon described them as "repugnant to the conscience of mankind," a phrase later incorporated into the Biological Weapons Convention.[38]

But, as Brad Roberts has written, "norms receive scant respect among foreign policy realists."[39] Notions such as abhorrence, repugnance, or morality are commonly ignored or belittled as possible instruments of policy. In reference to biological and chemical weapons, however, this perspective is regrettable. Reinforcing the attitude of abhorrence makes the development and use of these weapons less likely.

Bioweapons especially are seen as repugnant because they contravene an essential value among civilized people. Fighting illness is everywhere deemed a virtue. Deliberately to cause disease contradicts a deeply human sensibility. This translates into viewing biologicals as unacceptable instruments of war. Not just because of uncertain military utility, therefore, but because of human values, biological warfare has been rare. The prevailing attitude has been, in the words of biochemist Matthew Meselson, "You just don't do it!"[40]

But if prohibition of biologicals can be implemented with reasonable success, a next step seems possible. Demonstrating that one category of weapons is beyond the pale suggests the possibility that others might become pariah weapons as well. Indeed, the effort is underway for chemical weapons. And if possible with chemicals, why not eventually nuclear—all the weapons of mass destruction?

No less important than the details of arms control agreements is fostering a culture of antipathy. This does not minimize the imperatives of treaties, verification arrangements, and penalties for violators. But the likelihood of success is enhanced by reinforcing the sense of moral repugnance.

Underlying this book's recitation of facts and events is a profound human dilemma. It is the challenge of reconciling sometimes contradictory impulses between morality and survival. The book's three parts deal with conflict between moral behavior and national security. Part I expresses the tension involving biological and chemical warfare activities undertaken in the name of protecting the country. Army experiments and other government activities risked the health and safety of unwitting citizens. Some died as a result. These citizens, of course, were among the people the government was supposed to be protecting from an enemy.

Part II highlights how ethical compromise on the international level also led to unhappy irony. In failing to energetically repudiate Iraq's use of chemical weapons against Iran, countries sacrificed moral principle to other perceived interests. Then, in 1991, many of these countries had to face a biologically and chemically equipped Iraq on the battlefield.

Part III deals with new challenges, which have largely arisen from earlier expediential behavior by the United States and others. It addresses the increased threat of biological and chemical terrorism and the efforts to eliminate these weapons through treaties, invigorated norms, and other measures.

In the end, *The Eleventh Plague* is a demonstration of how the compromise of moral principles has led to greater *insecurity*. The book emphasizes the need to nurture the universal sense of revulsion about poison weapons. The ultimate benefit of an ethos that does not allow for biological or chemical warfare—for a man-made plague—is the example it provides. Extending such a mind-set to other weapons systems then becomes more feasible.

In the Name
of Defense

CHAPTER 2

OLD TESTS, FAILED ETHICS

A t family gatherings, the descendants of Edward Nevin often talk about his unexpected death in 1950. The cause was infection from a bacterium called *Serratia marcescens*. Nevin was one of 11 patients at Stanford University Hospital, then in San Francisco, who developed *Serratia* infections in a brief period. Illness from these bacteria had never before been recorded at the hospital.

A quarter century later, the family learned that a biological warfare test had been conducted in San Francisco about that time. Three days before the outbreak began, the army had sprayed the city with *Serratia marcescens*. Nevertheless, the government contended that the timing of the test and the infections was coincidental. The family sued for damages, but a federal judge in 1981 ruled in favor of the government.[1]

Unlike the Nevin family, plaintiffs won a court judgment for the 1952 death of Harold Blauer. A patient at a psychiatric hospital in New York City, Blauer received injections of a mescaline derivative prepared by the Army Chemical Corps. The drug was being investigated as a possible chemical weapon, and the army wanted to test it on a human subject. After the fifth in a weekly series of injections, according to the drug-study notes, Blauer "began sweating profusely and flailing his arms wildly . . . body stiffened . . . teeth clenched . . . frothing at the mouth." Two hours later he was dead.[2]

For Eric Olson, the 1953 death of his father Frank was "as if the lights went out." He recounted his family's four decades of anguish to a congressional committee in 1994. Frank Olson was a scientist at the army's biological warfare laboratories in Fort Detrick, Maryland. He swallowed a drink that, without his knowledge, had been spiked with lysergic acid (LSD) as part of a test by the Central Intelligence Agency. In a state of disorientation, he later plunged to his death from a New York City hotel room where he had been taken by a CIA agent.

Olson's family was told nothing about the LSD, only that he committed suicide. They learned about the drug and the CIA involvement in 1975 from a newspaper story. In 1994, the family had the body exhumed. A forensic specialist determined that the most likely cause of death was homicide.[3]

In 1994, residents in Minneapolis reacted furiously when they learned that they had been targets in an army experiment. News accounts reported that in 1953 the army sprayed the chemical zinc cadmium sulfide in simulated biological warfare attacks on the city. One of the targeted areas was the Clinton elementary school. Hundreds of former students now claimed they suffered inordinate amounts of illness over the years as a result of exposure to the chemical.[4] Congress funded a study to assess the health effects in Minneapolis and other cities where the chemical was sprayed. Results are expected at the end of 1996.

What moral value can there be to tests that risk the lives of citizens whose safety is the purported rationale of the tests? The question did not seem to interest the investigators who conducted the experiments.

At Senate hearings in 1977, the Pentagon acknowledged conducting 239 open air biological warfare tests over populated areas between 1949 and 1969. Locations included the cities of San Francisco, Minneapolis, St. Louis, Key West, and Panama City in Florida. Some tests were more focused, including the release of bacteria into the New York City subway system, into the Washington, D.C., National Airport terminal, and onto the Pennsylvania Turnpike.[5]

The tests were intended to see how bacteria might spread and survive in a biological warfare attack. Contending that the tests posed no risks, the army said that the bacteria and chemicals were harmless "simulants" of more lethal bioweapons. Accordingly, the army did not monitor the health of people in the test areas. It had "made an assumption of the innocence of these organisms," said a general at the 1977 hearings.[6] In fact, every one of the simulants was known at the time of spraying to be capable of causing illness or death.[7]

The four principal simulant agents were the fungus *Aspergillus fumigatus,* the chemical zinc cadmium sulfide, and the bacteria *Serratia marcescens* and *Bacillus subtilis.* In the years since experiments were conducted directly in cities, open air testing has continued at Dugway Proving Ground, 70 miles from Salt Lake City. But in tacit, if belated, recognition of their danger, the army no longer sprays the first three of these four simulants outdoors. The record on zinc cadmium sulfide is painfully instructive.

Operation LAC and Zinc Cadmium Sulfide

The Army Chemical Corps' *Summary of Major Events and Problems* for 1958 describes a series of tests called "Operation LAC."[8] The army literally had come to consider the entire United States an experimental laboratory. From the text:

> Operation LAC, which received its name from the initials of the words, "Large Area Coverage," was the largest test ever undertaken by the Chemical Corps. The test area covered the United States from the Rockies to the Atlantic, from Canada to the Gulf of Mexico. In brief, the Corps dropped a myriad of microscopic particles from a plane, and determined the distance and direction these particles traveled with the wind. The Corps wanted to learn these things: would it be feasible to contaminate a large area by this method using, for example, BW [biological warfare] organisms, and if so, what logistics would be involved.

The first test took place on 2 December 1957. A C119 "flying boxcar," loaned to the Corps by the Air Force, flew along a path leading from South Dakota to International Falls, Minnesota, dispersing fluorescent particles of zinc cadmium sulfide into the air. A large mass of cold air moving down from Canada carried particles along. Meteorologists expected the air mass to continue south across the United States, but instead it turned and went northeast, carrying the bulk of the material into Canada. The test was incomplete, but it was partially successful since some stations 1200 miles away in New York State detected the particles.[9]

Embedded in these paragraphs are stunning implications. The reference to the unprecedented scope of the test is un-abashed and represented as a "major event," not a "problem." Yet the observations by the unnamed author(s) of the 1958 *Summary* are replete with_problems. The zinc cadmium sulfide particles that the army dispersed posed health risks to the exposed popu-lation. Cadmium in particular was known at the time to be toxic and potentially carcinogenic, yet this fact is never acknowledged in the summary report. Indeed, cadmium oxide was a standard-ized chemical warfare agent during World War II.[10]

The army guessed wrong about meteorological conditions. Most of the material went off course into Canada. Partial "suc-cess," according to the report, meant that the uncontrolled par-ticles were identified 1200 miles from the points of release. Additional references to Operation LAC are equally unsettling:

> Dugway [Proving Ground in Utah] ran a second trial in February 1958. This time the "polar outbreak," as the Canadian cold air masses are generally called, continued on to the Gulf of Mexico, carrying fluorescent particles with it. As the air mass moved south the front broadened so that the line of particles 200 miles long at the aircraft's path spread out to 600 miles at the Gulf. . . .
> During the spring of 1958 Dugway conducted two additional tests, this time with the wind blowing haphazardly instead of steady [*sic*] from the north. In the first, the plane flew from Toledo, Ohio, and then turned

west to Abilene, Texas. In the second, the course ran from Detroit to Springfield, Illinois, then west to Goodland, Kansas. Sampling stations on both sides of the flight path reported particles, proving that random flight over a target area would disperse small particles widely.[11]

Thus, spraying was performed during steady weather conditions as well as when the wind was blowing "haphazardly." The first flight pattern indicated that the chemical particles were released over Ohio, Indiana, Illinois, Kentucky, Arkansas, Oklahoma, Louisiana, and Texas. The second pattern covered Michigan, Illinois, and Kansas. And these areas are just where the planes sprayed directly. The acknowledged dispersion of particles over hundreds of miles meant that most citizens in the nation may have been exposed to the army's experimental chemicals. What did the army learn from this exercise? The description in the 1958 *Summary* concludes:

> These tests proved the feasibility of covering large areas of a country with BW agents. Many scientists and officers believed this was possible, but LAC provided the first proof. While the tests were a great step forward, they did not provide the Corps with nearly as much data as the Corps would like to have had in order to predict the behavior of particles released in clouds. To obtain additional data the Corps planned further tests for the next fiscal year.[12]

The army's pleasure about the tests was unambiguous: "a great step forward." Having learned that a land area the size of the United States could be contaminated by biological agents, what else could the army learn from further outdoor spraying that would put the public at risk? Nothing more is said on the subject in any summaries through 1962, the last one available. We are left only with the promise in the 1958 *Summary* that further mass spraying was planned.

THE MECHANICS OF LAC

Additional information is available in a separate 1958 report from the biological warfare laboratories at Fort Detrick, Maryland.

Written by Herbert Beningson, the document explains the mechanics behind LAC. During a flight along a line of 400 miles, 5000 pounds of zinc cadmium sulfide would be released. The experiments entailed having a C119 cargo plane fly about 200 miles per hour, disseminating some 40 pounds of chemical particles per minute.[13]

In fiscal year 1958, the chemicals were released during flight time totaling "approximately 100 hours." This included "4 successful dissemination runs of varying distance including one of 1400 miles in length." No mention in the main body of the Fort Detrick report is made about the toxicity of the chemicals or their risk to the exposed population. The text concludes with a note of pride about the year-long operation: "During this time there has been no mechanical failure in any of the equipment."[14]

The Fort Detrick report contains two appendices, however. The first is a five-page (single-spaced) operation manual. The manual offers a detailed set of instructions. Its numbing narrative can be gleaned from a typical paragraph:

> Once in flight, it is well to make an occasional check of nuts, bolts and tie-downs, etc. Also, at some time before dissemination, the travel length of the drum unloaders should be adjusted. This will be determined by the dissemination rate and the amount of material packaged in each drum. A mark should be made on the ratchet wheel of each drum to indicate the notch in which the pawl rests when the drum and carriage are raised to the initial position. This position is determined by the unloader carriage so that the top of the material in the drum is one to two inches below the inlet of the feed duct. The drum should rise until its bottom is about one inch from the bottom of the feed duct to assure emptying the drum. The unloader stop should be set to trip the operating handle at this time. The stop is held in position by an allen set screw and the proper trip position can then be determined by experimentation.[15]

Although the mechanics of the operation were tediously recorded, the health risks were ignored. The only reference in the appendix to the condition of any persons appears in a single sentence: "One crew member will assist the others when necessary and act as spare in case of incapacitation or undue fatigue."[16]

The second appendix is untitled, but its two pages are unusually revealing. It notes that "experience with FP [zinc cadmium sulfide fluorescent particles] on actual trial operation had shown a high level of aircraft contamination caused by opening and handling the drums, spillage, residue after dissemination, etc." Nothing is said about the danger of these high exposure levels to the crew or whether special precautions were taken. But what follows is a surprising revelation, though some background information is necessary.

ACKNOWLEDGING RISK FROM *BACILLUS SUBTILIS*

For the past 50 years, the army has insisted that the simulant agents it releases outdoors pose no danger. Its most commonly used biological simulant is *Bacillus subtilis,* often referred to in army reports as BG (which stands for *Bacillus globigii,* another name for the bacteria). BG was used in tests directly over American cities and continues to be sprayed outdoors over Dugway Proving Ground in Utah.

Through the years, the Pentagon repeatedly has contended that BG is harmless. A 1977 army document provided to Congress maintained that "there is no evidence of infection in man or experimental animals following exposure to BG spores, even in massive doses."[17] An army spokesman in 1988 said: "We can find no documentation nor indication that the simulants used in outdoor testing [i.e., *Bacillus subtilis,* or BG] have ever been implicated as opportunistic pathogens."[18] A 1993 army statement on outdoor testing with *Bacillus subtilis* at Dugway Proving Ground noted: "No specific safety controls or protection are required for testing with simulants."[19]

Evidence has long existed in the medical literature that *Bacillus subtilis* can cause disease. Although not considered a widespread threat, the bacterium has been known to cause infections in people weakened by other conditions—in the very young, the very old, or people with immune deficiencies, in wounds following trauma, in surgery, or from prosthetic materials.[20] At no time has the army publicly acknowledged these risks. Yet Appendix B in the 1958 Fort Detrick report raises questions about what the army truly believed early in the program. The appendix reveals that the "ultimate aim was to allow BG and FP to be released alternately in a full-scale LAC trial." But, after finding that FP releases left high concentrations of the chemical in the aircraft, the army decided not to spray the bacteria as had been planned: "It was felt that a comparable level of contamination of BG within the aircraft would constitute a health hazard."[21]

True, concern was for the crew handling the materials, not the general public that might be exposed. But here was explicit acknowledgment by the army that concentrations of *Bacillus subtilis* constituted a health hazard to people exposed. This was in 1958. Spraying these bacteria in other tests in various parts of the country continued, including into the New York City subway system in 1966. Even less credible now is the army's claim that it believed its tests could cause no harm.

But BG aside, what did the army know about the hazard of exposure to zinc cadmium sulfide? As revealed in the 1977 Senate hearing on biological warfare testing, fluorescent particles, designated by the army as "FP," had been used over dozens of populated areas between 1949 and 1969. The FP was zinc cadmium sulfide. A list of targets indicates that the chemical was typically aimed at discrete locations, usually cities. Among them were Minneapolis, St. Louis, the San Francisco Bay area, Cape Kennedy in Florida, Fort Wayne, and Cambridge, Maryland. The list also mentions "Continental U.S. East of Rocky" [*sic*] in tests conducted during 1957 and 1958.[22] Not until the 1990s, after information about the tests was declassified, could the description "Continental U.S." be understood as literally true.

THE ZINC CADMIUM SULFIDE HAZARD

At the outset of the testing program, the army had contracted Stanford University chemist Philip A. Leighton to help develop a tracer for open air biological and chemical experiments.

Leighton described the rationale for using zinc cadmium sulfide in a 1955 book, *The Stanford Fluorescent-Particle Tracer Technique.*[23] By then the material had been used in several outdoor biological warfare tests, including in Minneapolis and St. Louis in 1953. Inasmuch as most of the American public may have been breathing in the army's zinc cadmium sulfide during Operation LAC, an essential question is what the army really knew about its risks.

Although Leighton's book reveals his contractual relation with the army, it says nothing about the tests that were then being secretly conducted. Only after the public learned about the testing program in the 1970s was Leighton's work acknowledged as the basis for the army's use of the chemical. The book refers to zinc cadmium sulfide as No. 2266 FP. This was the name given by its manufacturer, the New Jersey Zinc Company.

The chemical mix was used because the size of the particles, their stability, and their dispersibility simulated biological warfare agents. The particles would accumulate on special filters at various distances from the point of release. Because they glowed when viewed under ultraviolet, or "black," light, the accumulated particles on the filters could be counted. In this manner, the concentrations and effectiveness of the mock biological agents might then be gauged.

Leighton's 1504-page book, subtitled *An Operational Manual,* contains two pages on the toxicity of zinc cadmium sulfide. His initial observations are hardly comforting:

> Compounds of zinc and cadmium are both known to be poisonous when taken into the human system. For this reason, the No. 2266 FP material is labeled with a poison warning when shipped by the manufacturer, and in applications of the material the possibilities of toxic effects must be considered. Cadmium is the more poisonous of the two metals, and since it is very likely that each acts independently of the other, the toxicity of the cadmium is the more important factor.[24]

Leighton then contends that the estimated quantities to which people will be exposed are small and have no toxic effects.

How does he know? He was told this by Herbert E. Stokinger of the U.S. Public Health Service "in a private communication concerning the hazards of breathing the FP material." Leighton footnotes four studies on cadmium poisoning to justify his contention, indicating that they were "references cited by Dr. Stokinger."[25]

There is no indication that Leighton searched the scientific literature independently or even read the four studies cited by Stokinger. The first of those studies concluded that its "data suggest that cadmium may not be as great an industrial hazard as has previously been reported."[26] The second did not detect pulmonary edema associated with acute cadmium poisoning among five men exposed to cadmium in their work. It concluded, however, that "these men may be developing such pulmonary damage slowly, if exposure to cadmium above a now unknown level persists."[27]

The third study found urinary pathology in workers who had been exposed to cadmium and nickel dust for at least eight years, but not among workers who had been exposed "during a period of a year or two."[28] The fourth investigation found no deleterious effects from inhalation of cadmium at low levels but acknowledged that "other investigators have demonstrated a definite effect . . . particularly on the red blood cells."[29]

One can fairly conclude from these four studies that the evidence about low-level cadmium toxicity was unclear. To cite them as confirmation of safety for the kind of work the army was doing is plainly unwarranted. The allowable maximum concentration of cadmium for workers was 0.1 milligram per cubic meter. No one knows the concentrations that citizens were being exposed to during the army tests or the exposure levels to crew members who were releasing the material.

Perhaps most disturbing is the absence of references in Leighton's book to studies that found serious health risks from low-level cadmium exposure. One of the earliest appeared in a 1932 issue of *The Journal of Industrial Hygiene*. In it, Leon Prodan showed that "cadmium is a dangerous substance and that the type of damage to be expected is of such critical nature as to indicate the avoidance of the inhalation or ingestion of even small amounts of cadmium." Prodan demonstrated that very small amounts of cadmium oxide fume or dust, as well as of cadmium sulfide, may produce serious damage in the lungs. On the basis of these observations one may conclude that cadmium, no matter how small the amount taken in the lungs, causes pathologic changes, and that there is, therefore, no permissible amount of cadmium.[30]

The findings and recommendations are so categorical that they surely should have raised concerns about outdoor spraying. Prodan's study is not even cited in Leighton's manual, let alone its conclusions acknowledged. Like Leighton, the army either ignored or misrepresented findings that suggested its chemical was unsafe. It continued to spray zinc cadmium sulfide in outdoor tests until the early 1970s.

The chemical was taken out of use, apparently in reaction to efforts by L. Arthur Spomer, now a professor in the School of Agriculture at the University of Illinois. Spomer was a young army officer working as a meteorologist at a military base in Salt Lake City in the early 1970s. During a trip to Dugway, he observed soldiers preparing a mix of zinc cadmium sulfide for open air spraying. He told his superiors about the potential dangers of the material but initially was ignored.[31] After reviewing the scientific literature, he published an article in 1973 showing that, since the early 1930s, cadmium was "known to be toxic to almost all physiological systems."[32]

Spomer's commentary is an indictment of the entire outdoor spraying program:

> Although Cd [cadmium] toxicity is well-established and
> FP [zinc cadmium sulfide fluorescent particles] is
> commonly used as a tracer in atmospheric studies,
> no case of Cd poisoning resulting from the use of FP
> has been reported in the literature. This may be because
> none has occurred; however, it is more likely that
> such poisoning has been of a low-level chronic nature
> and its symptoms are less dramatic and more difficult
> to recognize than in the case of acute Cd poisoning.
> A general ignorance of the toxicity of FP and of the

symptoms of Cd poisoning also contribute to the failure to recognize FP poisoning.[33]

The appearance of Spomer's article in the open literature undoubtedly influenced the army to reconsider its use of the chemical. Although the army stopped spraying zinc cadmium sulfide outdoors, it continued to maintain that the tests harmed no one. In 1980, an army spokesman said, "Ingesting zinc cadmium sulfide is like swallowing a pebble. It is a nonsoluble material. It would pass through you."[34]

In 1994, Senator Paul Wellstone requested information about the risks to Minneapolis residents who had been exposed to the chemical in 1953. The army produced a 29-page risk assessment, acknowledging that exposure to cadmium compounds could cause health problems. But it concluded that the estimated exposure levels in the 1953 tests were low enough that they "should not have posed any adverse health effects for residents in the test areas."[35] The last three pages of the 1994 assessment list 38 references. Only 12 are related to cadmium toxicity. Entirely absent are works like those of Prodan or Spomer that cast doubt on the army's claims of safety.[36]

Yellow Fever

Beside the spraying of most of the continental United States with toxic chemicals, other disquieting tests were revealed in the Army Chemical Corps's annual summary reports. The 1959 *Summary* noted that in 1953 the biological warfare laboratories at Fort Detrick began to study how insects might be used to spread disease-causing agents. Insects and other arthropods offer advantages, according to the report, because "they inject the agent directly into the body, so that a mask is no protection to a soldier, and they will remain alive for some time, keeping an area constantly dangerous."[37]

One army favorite in this regard was yellow fever, described in the report as "a highly dangerous disease." The yellow fever virus is carried by a mosquito called *Aedes aegypti*. In the course of biting and sucking the blood of people or animals, the female mosquito injects them with the virus. Conversely, an uninfected mosquito can become infected if the person or animal it bites is carrying the virus.

A few days after being bitten, victims develop fever and headaches, begin to vomit, and suffer from prostration. The result can be either death or recovery that takes months. The report notes further that "there is no known therapy for yellow fever, other than symptomatic, and in severe cases the patient has a poor chance of recovering." Coupled with the presumption that a military attack with mosquitoes would be difficult to detect, the army report called "the *Aedes aegypti*–yellow fever combination an extremely effective BW agent."[38]

Midway through its rationale for using yellow fever as a warfare agent, the report recounts several unusual experiments to test "the practicality of employing *Aedes aegypti* mosquitoes to carry a BW agent." From the text:

> In April–November 1956 the Corps ran trials in Savannah, Georgia, by releasing uninfected female mosquitoes in a residential area, and then, with the co-operation of people in the neighborhood, estimating how many mosquitoes entered houses and bit people.[39]

The report does not indicate how many mosquitoes were released during the seven-month period. Nor does it say how many people were bitten, though apparently some "estimated" number were. The comment about people's cooperation seems limited to their telling test officials in some unspecified manner that they had been bitten. Savannah residents evidently were unaware that they were part of an experiment. Nothing is said about seeking informed consent from the targeted people. One can only wonder whether they would have agreed to being exposed to mosquitoes that can carry a deadly virus, even though the army proclaimed the mosquitoes to be uninfected.

The mosquito tests were not limited to Georgia. The text continues:

> Also in 1956 the Corps released 600,000 uninfected mosquitoes from a plane at Avon Park Bombing Range,

Florida. Within a day the mosquitoes had spread a distance of between one and two miles and had bitten many people. In 1958 further tests at Avon Park AFB [Air Force Base], Florida, showed that mosquitoes could easily be disseminated from helicopters, would spread more than a mile in each direction, and would enter all types of buildings. These tests showed that mosquitoes could be spread over areas of several square miles by means of devices dropped from planes or set up on the ground.[40]

Thus, between 1956 and 1958, the army repeatedly released mosquitoes from airplanes, helicopters, and ground devices. The passage discloses that "many" unsuspecting citizens were bitten. They had become test subjects without their knowledge or consent. Apart from ethical questions, how dangerous to the general population were these experiments? According to Dr. Barry Miller, an entomologist with the Centers for Disease Control and Prevention in Fort Collins, Colorado, the risk was small. But it could not be entirely dismissed. *Aedes aegypti* mosquitoes, he says, existed in that part of the country before the army experiments. Some may well have carried the yellow fever virus, and adding more mosquitoes would increase the pool of potential carriers. The newly introduced females could become infected if they bit animals or people who were infected already.[41]

John P. Woodall, director of the arbovirus laboratory for the New York State Department of Health, called the army's experiment a "terrible idea." (Arboviruses are viruses carried by mosquitoes or other arthropods.) Woodall, who previously worked with the World Health Organization, worried because *Aedes aegypti* also transmit other pathogenic viruses, notably those that cause dengue fever.

The released mosquitoes could have established their own breeding areas. If even a few males were among the females, Woodall conjectured, the mosquitoes might create a niche in the environment and thrive there indefinitely.[42] Clearly, the army's experiment increased the health risk to the human population.

Ignoring the safety and ethical questions, the army was pleased with its test results. As a consequence, it planned a new entomological production facility. A goal of the facility would be to produce 30 million infected mosquitoes every week.[43]

K-Agents

Among the most disturbing references in the Chemical Corps' *Summaries* are those to psychochemical agents. For such an agent, "it was desirable, but not essential, that it have no permanent effect."[44] Sometimes referred to as K-agents, psychochemicals cause "stupefaction, hallucinations, incoordination, faulty judgment, depression, or other types of mental incapacitation." The 1957 *Summary* names three types of K-agents: mescaline derivatives, lysergic acid (LSD), and tetrahydrocannabinol derivatives (marijuana-related chemicals).[45]

The text indicates that in 1951 the army awarded a contract for the study of mescaline derivatives to the New York State Psychiatric Institute. Based on the work there and in its chemical warfare laboratories, the army decided that "mescaline and its derivatives would not be practical as an agent because the doses needed to bring about the mental effects were too large."[46] The benign wording obscures the story of how its experiments killed Harold Blauer in 1953 while a patient at the institute.

The 1957 *Summary* indicates that in addition to mescaline, the New York State Psychiatric Institute was investigating the effects of LSD. Arrangements with Tulane University and the University of Maryland Psychiatric Institute were noted as well— also to determine whether LSD might be "an effective chemical warfare agent." The hope was that the drug would disorient an enemy. Soldiers might then disobey their commanders and become unable to operate tanks, missile launchers, and other equipment. Nowhere is there an indication of how the experiments were conducted. The human subjects, according to the report, were army volunteers.[47]

The report's final paragraph with reference to K-agents is especially disquieting. It says that the army was sponsoring an investigation at the University of Michigan on the effects of a tetrahydrocannabinol derivative. The experimental subjects were cats and dogs, but there is no description of the test.

Without elaboration, the paragraph concludes: "Preliminary observations suggest that the compound may bring on pathological changes, and if this proves to be true, the testing of the compounds on volunteers will have to be done with great caution."[48]

The sentence raises intriguing questions. Why single this experiment out as needing "great caution"? Were not *all* the experiments with these potentially dangerous chemicals done with great caution? What specifically are the pathological changes referred to in the document? Who would the human "volunteers" be, and what would they be told about the pathological changes they might suffer? If the risk is as foreboding as the report suggests, should human subjects be used at all?

By the time this *Summary* was issued, both the army and the CIA had dispensed mind-altering drugs to unsuspecting victims. Harold Blauer had already been killed by a mescaline-derivative experiment. Frank Olson underwent psychotic changes after LSD had been planted in his drink by a CIA agent. As a result he leaped, or was pushed, to his death through a hotel room window. Not a hint of these or other terrible consequences of the mind-altering experiments is mentioned in any of the reports.

Quite the contrary. A reference to K-agents in the 1958 *Summary* is rather sunny. The effects are characterized as pleasant and euphoric. The report described an experiment with an LSD derivative on "human volunteers":

> One of the most interesting investigations was carried out to see if the drug would affect a squad of men who were undergoing routine training. The leader, however, did not receive the compound. The men paid little attention to their leader's commands. Their motions were slow, they were quite happy and unconcerned with the leader's attempt to drill them. In the second experiment the squad leader as well as the men received the drug. When an officer told the leader to drill the men he refused and told the officer to do it himself. The group laughed and joked. When the officer told the squad leader to leave the field, he refused and had to be escorted away.[49]

The report concluded that K-agents might have great military and even humane value: "Possibly entire enemy positions or forces could be subjected to the K-agents and captured without resistance or casualties."[50] The conclusion sounds appealing. There is no suggestion of long-term pathological or other harmful effects on the "volunteers." Whether or not the subjects experienced problems from the tests is not mentioned. How long the subjects were monitored after the tests, if at all, is not discussed.

In fact, the tests were conducted according to an advisory that informed consent be disregarded. A committee under authority of the assistant secretary of defense for research and development in 1956 urged that subjects in LSD tests not be informed about what they really were participating in:

> In view of the fact that a great many of the effects observed
> in the group may be the result of suggestion (placebo
> effect) it would appear desirable to have one control
> group which has neither been given a training lecture
> on LSD-25, nor any information as to the symptoms
> of the drug being administered.[51]

Not until 1975 did the army question this approach. After reviewing the overall experimentation program, the army's inspector general was frankly critical: "What is involved is that in spite of clear guidelines concerning the necessity for 'informed consent,' there was a willingness to dilute and in some cases negate the intent of the policy."[52] Nevertheless, the dreadful effects of other mind-altering tests sponsored by the army or the CIA were never publicly acknowledged by either institution.

Operation Whitecoat and CD-22

"The ultimate objective of Operation 'CD-22'," according to a 77-page description of a 1955 experiment, "was to effect the exposure of human test subjects to a typical BW aerosol." Obtained in response to a 1994 Freedom of Information Act request, the CD-22 report describes an experiment that is as weird as any in the army's biological program. The aim was to determine whether, under field conditions, people were susceptible to infection from an aerosol of *Rickettsia burnetii*.

Rickettsiae are classified between bacteria and viruses because they share characteristics with both. The organisms used in the test, *Rickettsia burnetii* (also called *Coxiella burnetii*), were released from generators 3200 feet upwind from the target areas. Initial experiments were on guinea pigs and monkeys. Later, people were included.[53]

The abstract of the study concludes that "guinea pigs, monkeys, and man exposed concurrently to the same estimated dose-levels show similar responses."[54] The report's introduction makes clear that Operation CD-22, also termed Operation Whitecoat, purposely employed actual biological warfare agents. No simulants here—no organisms or chemicals that the army would even try to pretend were harmless.

Thus, biological agents that the United States planned to use against an enemy were now being aimed at Americans. Although the spraying took place over Dugway Proving Ground, once the agents were released into the air, no one could control their path. The vagaries of meteorological conditions became clear in the report. At one point, for example, aerosol generators that had been placed along a designated line were moved so that the operation would benefit from the "prevailing wind." Elsewhere the report notes: "Estimated dosages achieved at positions closest to the generator source varied considerably from trial to trial. These variations can be attributed, in part, to differences in meteorologic conditions."[55]

These rickettsiae cause Q fever; its symptoms include pneumonitis, fever, and vomiting. The army report claims that prognosis for recovery is excellent, especially when therapy is prompt.[56]

The army's confidence about safety and recovery is not matched by descriptions in the medical literature of the time. According to a standard 1952 medical textbook: "Few fatal cases of Q fever have been reported, and, in most of these, autopsies were not performed. Consequently, satisfactory data are not available for general remarks on the pathology of Q fever." The disease is susceptible to antibiotic therapy, the textbook continues, and recovery is usually complete: "but relapses have occurred in several instances; they are similar to the primary disease and may be mild or severe."[57]

A clinically dry paragraph in the army report describes the placement of the animals and people in preparation for the spraying:

> A total of 75 rhesus monkeys and 300 guinea pigs were utilized in the Phase IV trial. The monkeys were positioned, in groups of either seven or eight, at 10 stations, H 22 through 31, near the center of Row H, while the guinea pigs were emplaced at 30 stations, H 12 through 41, along the row in groups of either 5 or 20. Thirty human volunteers participated in the test. One human protected by previous immunization against the agent and two who were not so immunized were emplaced at each of the Stations H 22 through 31. All volunteers were provided with chairs of an appropriate height to accomplish sampling at 5 feet above terrain.[58]

As if in a huge theater, locations for the audience were designated by rows hundreds of feet apart. Members of the audience sat in assigned seats while the performers—trillions of microorganisms—were being prepared off-stage for entry. The people were propped in elevated chairs, high enough to enhance the likelihood that they would be inhaling the biological agents. Seating accommodations for the animals were unspecified. But as the show began, all were engulfed in clouds of infectious agents. The line between spectator and performer disappeared as all became participants in a benighted drama.

The animals became infected, though data about the human effects were "not available for inclusion in this report." But the document concludes with a suggestive peek. While noting that the human data would be detailed elsewhere, it said:

> Word has been received, however, that positively diagnosed infections occurred in some of the human volunteers who were positioned at stations where the greatest agent dosages were achieved, and that these responses paralleled those observed in the monkeys and guinea pigs at the same stations in this trial.[59]

Because documentation about illness in the human subjects was unavailable, what precautions or therapy were instituted are unclear. Subsequent publications about the overall testing program suggest the infected human subjects received antibiotic therapy. Moreover, before agreeing to participate, the subjects may have been informed about the nature of the test, its dangers, and its health risks. But that is uncertain. As previously noted, the army's 1975 inquiry into the use of human subjects in biological and chemical warfare research reported failure to abide by the requirement of informed consent.[60] Even less likely than the subjects' receiving appropriate information is that their long-term health was monitored. For Q fever, that is particularly significant. According to the medical text cited earlier, relapses are not uncommon.

The aerosol tests in 1955 were among the first that explicitly used human subjects in the biological warfare program. But the program continued for years. The 1959 *Summary of Major Events and Problems* tells of six tests over Dugway Proving Ground in Utah in which microorganisms were released from airplanes: "While five trials were made with simulants, one flight was made carrying the agent that causes Q fever."[61]

How effective was the airplane spraying? "Guinea pigs were placed at the sampling stations. Results indicated that if human beings had been in the area, 99 percent of them would have been infected."[62] The danger was not confined to the target area, however. If even a few bacteria had blown off course, they would have been dangerous. As an officer in the biological warfare command wrote, "Tularemia and Q fever can be produced in man by the inhalation of fewer than 10 organisms."[63]

The uncertain path of agents released into the air was underscored in the army tests with zinc cadmium sulfide. But a paragraph immediately following mention of the Q fever test emphasizes how the spray from a single airplane could range over thousands of miles. The tests proved that a "spray system could contaminate 50,000 square miles with BW aerosol in a single sortie. . . . [T]hree large aircraft, each carrying 4,000 gallons of liquid BW agent, and flying at a speed of 500 knots, could spray an area of 150,000 square miles, causing more than half the people in the area to become ill."[64] Nothing appears in the report about the danger of the Dugway tests to people outside the base.

Failed Ethics

A preeminent concern about the ethics of the army's research is the matter of informed consent. The army does not consider spraying simulants over populated areas to be research implicating human subjects. Its rationale is that no one in particular is being targeted. Individuals who unwittingly inhale the test agents just happen to be there, according to the army's thinking, and there is no requirement to obtain their consent.[65]

This line of justification ignores the rights of people who may not wish to participate. All the more is this true when the army intentionally tries to infect people or alter their mental states. The Nuremberg code places the requirement of informed consent beyond dispute. The 1947 code was part of the verdict rendered at the Nuremberg trial of Nazi doctors and officials who conducted research on involuntary subjects. Thousands of Jews and other concentration camp inmates were subjected to horrifying experiments that commonly ended in pain, disfigurement, and death. The code essentially enshrines the requirement that people be fully informed and give consent before becoming test subjects. How has the army dealt with the requirement?

Upon issuance, the Nuremberg code implicitly became the standard for human subject research in all civilized societies. It was explicitly adopted by the U.S. Department of Defense in 1953. A February 1953 memorandum from the secretary of defense to the secretaries of the army, navy, and air force dealt with the use of human volunteers in experimental research. In recognizing the need for human subjects in research related to biological, chemical, and atomic warfare, the memorandum incorporated the words of the code. Beginning with the admonition that "the voluntary consent of the human subject is absolutely essential," the memorandum, like the Nuremberg code, continues:

This means that the person involved should have legal capacity to give consent; should be so situated as to be able to exercise free power of choice, without the intervention of any element of force, fraud, deceit, duress, over-reaching, or other ulterior form of constraint or coercion; and should have sufficient knowledge and comprehension of the elements of the subject matter involved as to enable him to make an understanding and enlightened decision. This latter element requires that before the acceptance of an affirmative decision by the experimental subject there should be made known to him the nature, duration, and purpose of the experiment; the method and means by which it is to be conducted; all inconveniences and hazards reasonably to be expected; and the effects upon his health or person which may possibly come from his participation in the experiment.[66]

The document then restates the remainder of the Nuremberg code, including the requirement that the experiment be intended to "yield fruitful results for the good of society, unprocurable by other methods or means of study"; that it should "avoid all unnecessary physical and mental suffering and injury"; that it be "conducted only by scientifically qualified persons"; that a subject be permitted to "bring the experiment to an end" if it is causing him undue physical or mental stress.

Soon after, the secretary of the army made clear that the Nuremberg code would apply as well to firms and institutions under army contract. There should be care, he wrote in a memorandum, "that the same basic principles and safeguards applicable to the department of the Army laboratories are observed by the contractor."[67]

For reasons that are not clear, both memoranda were originally classified. In effect, however, the secret classification should have meant nothing. A note appended to the memorandum concerning contractors indicates that the code of ethics was to be "downgraded from Secret to Unclassified when separated from Secret inclosures."[68]

The initial memorandum from the secretary of defense, which recited the Nuremberg code, formally remained classified until 1975. But in June 1953 the army's chief of staff had directed the Chemical Corps to implement its provisions. Moreover, in 1954 this directive to the Chemical Corps was made unclassified.[69]

Thus, implicitly since 1947 and explicitly since 1953, U.S. biological and chemical warfare testers were under orders to enforce the standards of the Nuremberg code. The notion that ethical standards for human subject research in the 1950s were different from those of today is untrue, despite latter-day allegations to the contrary. The sensitivity of army researchers may have been different in the 1950s from what it is now. Levels of adherence to the rules may have been different. But the right of people to be fully informed and grant consent before becoming experimental subjects was as much an official mandate in the 1950s as in the 1990s.

In his 1975 review of the use of human subjects in Chemical Corps research, the army's inspector general is candid on the matter:

> Throughout the initial years of the Army volunteer program, the Chief of Staff's memorandum . . . was cited at all echelons as the "basic policy." Thus, there is little doubt that commanders and investigators involved in the use of volunteers in research had access to the Secretary of Defense instructions regarding the essentiality of informed consent for all volunteer subjects. Nevertheless, there was evidence developed during this inquiry to indicate that in many cases consent was relegated to a simple, all-purpose statement to be signed by the volunteer.[70]

The inspector general's report indicated conscious efforts to circumvent the requirement for giving full information to subjects. A consent form drafted in 1954 would have volunteers certify that the nature of the experiment as related to health hazards had been "explained fully." The form adopted for actual use omitted the word "fully." The volunteer would be told only about "the general nature of the experiments" as related to possible health hazards.

Also deleted from the proposed version was an entire sentence that read: "I certify that I have been familiarized with the nature of the experiment and the agents to be used, commensurate with security requirements."[71] A sample form was included in another compilation of material about the army's research on human subjects. Titled "Consent Statement," the text reads:

> I, _____, without duress and of my own free will do hereby consent to participate in a research study conducted by physicians of the U.S. Army Medical Research Institute of Infectious Diseases, Fort Detrick, Maryland, involving _____[two lines are left blank]_____. The implications of such a study have been explained to me. I understand that an element of risk is involved in this procedure. I understand that this is an approved research study and as such will be recorded in official files of the Department of the Army. Any medical problems arising from my participation in this study will be considered to have been incurred in line of duty. I also understand that no additional rights against the government will accrue from my having participated as a volunteer.[72]

Beside spaces for the signatures of the volunteer and two witnesses, that is the whole consent statement. The wording suggests less concern that the volunteer be properly informed than that the army be absolved of responsibility for problems that might occur.

The inspector general's 1975 review found that volunteers largely were not being informed about agents to which they would be exposed. The army had engaged in "selective compliance" with the rules, according to the review. This was attributable to "a frame of mind and purpose which fostered a willingness to bend or break rules and policies so as to insure mission accomplishment; a continuing reliance on the 'end justifies the means.'" The review concluded that "the intent of the informed consent policy did not appear to have been fulfilled."[73]

The review noted, however, that a revised 1975 consent form seemed an improvement. Included in the new form was the statement that

> The implications of my voluntary participation; the nature, duration and purpose, the methods and means by which it is to be conducted; and the inconveniences and hazards

which may reasonably be expected have been explained to me by _____, and are set forth on the reverse side of this Agreement, which I have initialed.[74]

The warrant for confidence that this new wording would improve protection of a test subject was conjectural. Questions continued afterward about risk and safety—including thoroughness of explanations about the experiment; the ability of a test subject to comprehend; and the influence that rewards or punishments might have on decisions to volunteer. The review recognizes that the manner of persuading someone to volunteer can be inappropriate. It reported a list of rewards that soldiers who volunteered as subjects had received: money, weekend passes, improved living and recreational conditions, relief from fatigue-type details, opportunity to receive a medical examination not available during other military assignments, and a letter of commendation.

None of these privileges was illegal. But in recognizing such attempts to persuade soldiers to volunteer, the review is mildly critical. The methods "appeared not to have been in accord with the intent of Department of the Army policies governing use of volunteers in research."[75] By emphasizing that these practices took place in the 1950s and 1960s, the review implied that, by 1975, they no longer were in vogue.

The overall biological defense program was dramatically reduced in the 1970s, including its human subject component. In the 1980s and 1990s, however, the program again began to grow. So did concerns about its safety and ethics.

Is It Safe?

Controls over "all phases of the BDRP [Biological Defense Research Program] . . . have effectively eliminated significant adverse impacts to the biophysical environment and to human health," the army declared in 1993.[1] No danger to anyone!

The assurances reiterated earlier claims prompted by a federal court action. In 1986, the Foundation on Economic Trends, an organization critical of genetic engineering, had filed a suit that questioned the safety of the BDRP. In consequence, the Pentagon agreed to prepare an environmental impact statement.

The "Final" Word

The army released an 875-page final programmatic environmental impact statement on the BDRP in 1989. It concluded that existing controls provide "adequate protection for the workforce and virtually total protection for the external environment." The program was "thoroughly analyzed," and "additional mitigation was not found to be justified."[2]

The Army's Contentions

According to the statement, conditions had never been safer at Fort Detrick, Maryland, which has been a principal location for military biological research since 1943. During the time that offensive work was being done there, through 1969, 419 incidents of illness were associated with the program. In the 1970s and 1980s, under the defensive program, only five people reportedly became ill. The statement contends that no member of the public had ever been infected by a Fort Detrick experiment. It says further that only three workers there ever died from infections: two from anthrax in 1951 and 1958, and a third from Bolivian hemorrhagic fever in 1964.[3]

These contentions are marred, however, by incomplete information. For example, Charles Dasey, a spokesman for Fort Detrick, acknowledged elsewhere that a janitor at the base died from anthrax infection in 1968. The victim was exposed while changing a light bulb in a building that was contaminated with the bacteria.[4] Omitting this fatality from the army's tabulation raises questions about how many others may be missing. Moreover, the location was only one of several "hot spots" at Fort Detrick and at Dugway Proving Ground in Utah, some of which remain contaminated.[5] None is mentioned in the army's environmental impact statement.

Federal law requires that public comments be considered in the preparation of an environmental impact statement. Thus, the army's 1989 final programmatic environmental impact statement was preceded by a draft in 1988 that invited comments from the public.[6] The final statement and the draft were virtually identical, except the final version included a 189-page appendix that summarized the comments and the army's responses.

The comments, whose sources were not identified, revealed concerns about accidental releases of pathogens during shipment or in the laboratory, outdoor aerosol testing, security, the threat of novel organisms, and the possibility of defensive work slipping into offensive. The army's response to each comment invariably sought to demonstrate that concerns were unwarranted.

Those who worried about widespread infections from the army's research, for example, could read that an "epidemic (i.e., a spread from person to person) resulting from organisms studied in the BDRP is technically and epidemiologically impossible."[7] Those concerned about accidents were referred to a section headed "Unexpected External Event." The section said in part:

> No plausible combination of human error or mechanical failures can be conceived that would result in materials being released because of design and redundancy of control systems, safety procedures, and mitigating and monitoring steps.[8]

The phrases were precisely the same in both the draft and the final statement, as were the concluding words of assurance in each—protection is adequate or total, and additional mitigation is unnecessary.

CONGRESSIONAL AND OTHER SKEPTICISM

Yet between the time of the draft and final environmental statements, a congressional committee held hearings on the subject. In July 1988, witnesses before the Senate Subcommittee on Oversight of Government Management gave testimony that contrasted sharply with the army's contentions. The hearings were prompted by a report that subcommittee staff members had issued in May. An 18-month review found "serious failings" in safety management of the Pentagon's chemical and biological research programs.[9]

In his opening statement at the hearings, subcommittee chairman Carl Levin indicated that "there is a disturbing record of safety problems at chemical research facilities" and that "the biological side has been in even worse shape." Among the problems with the biological program:

> There has been no readily identifiable organizational structure within DOD [Department of Defense] for overseeing safety; contractor facilities were not prescreened; there was a confusing and inadequate patchwork system of safety regulations, and no DOD safety inspections.[10]

Witnesses pointed out dangers associated with research on pathogens, whether for military purposes or not. Some who worked in the BDRP testified about health and safety failings at their facilities including unreported fires, accidents, and missing viruses and other biological materials.

The emergent sense about the overall program was suggested by a 1988 General Accounting Office (GAO) report—uncertainty. Levin's subcommittee had requested the report to assess safety in the chemical and biological programs. Even if protective practices in the biological areas were adequate, the GAO report said, no one could be sure. That is because "in the biological defense program, DOD has not developed its own safeguard requirements or conducted regular, formal evaluations of contractor facilities." The report recommended that a system of centralized evaluations for contractors be established. Meanwhile, in the absence of an effective evaluation process, "uncertainties will persist about the adequacy of existing safeguards governing biological research and development."[11]

Three years later, the GAO leveled new criticisms at the Biological Defense Research Program. In a report published in December 1990, it suggested that the army was "unnecessarily" duplicating medical research underway at civilian agencies. Moreover, only 112 of the BDRP's 218 medical research projects could be confirmed as being directed at biological agents that the army had "validated" as biological warfare threats. Conversely, several validated agents were not part of any research project.[12] Although the newer report did not revisit the safety issue, the GAO's findings fueled further skepticism about management practices in the BDRP.

In February 1991, the federal Occupational Safety and Health Administration (OSHA) added to the disquiet about the program's safety. The agency determined that the army's safety inspection program of biological defense research activities "did not meet many of OSHA's inspection requirements."[13]

Meanwhile, in January 1991, the army had announced the establishment of a new Biological Defense Safety Program and issued a pamphlet outlining the program's "technical safety requirements."[14] According to William Wortley, the army's designated contact person on the matter, the effort was in response to public concerns. Although no new safety practices were listed, the pamphlet was "a conglomeration of what was going on." The army's laboratories and those of outside contractors "had been doing everything correctly," he said, but "we wanted to put it all under a single control."[15]

The army's declarations about the safety of its program have been challenged from many quarters—Congress, government agencies, outside analysts. The army's credibility is hardly enhanced by gratuitous assurances, as in its 1989 final programmatic environmental impact statement, that controls provide "virtually total protection" of the environment, that error and failure cannot be "conceived," that an epidemic is "impossible." Such exaggerations ignore the fallibility of human activity and raise questions about the value of the army's assessments.

Four years after the army's "final" assurances about safety, the program was being questioned as forcefully as ever. The Center for Public Integrity released a scathing critique in 1993. Titled *Biohazard: How the Pentagon's Biological Warfare Research Program Defeats Its Own Goals,* the study concluded that the program was wasteful, provocative, and "consistently failed to proceed in a fully open and accountable manner."[16]

The Levitt Case

One of the witnesses at the 1988 Senate subcommittee hearing was Neil Levitt, who had worked at the U.S. Army Medical Research Institute of Infectious Diseases (USAMRIID) in Fort Detrick. He started there in 1969 after receiving a Ph.D. in microbiology. After three years as an army officer, he stayed on as a civilian research scientist. He left in 1986 at the age of 44.

THE MISSING VIRUS

In 1979, Levitt was asked to develop a vaccine against the Chikungunya virus. Considered a biological warfare threat, the virus causes fever, malaise, headaches, and severe muscle and joint pain. The intention was to grow an attenuated form of the organism. If a person were inoculated with a weakened virus, it was hoped he would develop antibodies that would also resist the virulent form. Levitt was instructed to grow the virus in a special cell line called fetal rhesus lung cells. (The cells were originally derived from the lungs of fetal rhesus monkeys.)

The army had previously tried to develop other vaccines by using the same cell line to grow weakened viruses. It had been necessary to abandon those efforts, however, because the viruses reverted to lethal form after being inoculated into human subjects.[17]

Tests by Levitt and his assistant suggested that similar problems were occurring with the Chikungunya virus. Two-thirds of the experimental mice died after receiving injections of fluid from those cells. Levitt informed his superiors about the problem and recommended in USAMRIID's 1981 annual fiscal report that the fetal rhesus cell line not be used for human vaccine production. He was surprised to find his recommendations deleted from the report sent to Congress: "Quite the contrary, only a glowing report of the vaccine's progress was discussed." He refused at that point to do further work with the cells toward vaccine production.[18]

The stock of Chikungunya virus that had been grown in the suspicious cell line was then stored in a freezer at the institute. In a routine inventory in September 1981, Levitt discovered that the entire stock was missing. He reported this to his department and division chiefs. They confirmed the disappearance of some 60 vials containing about 2500 milliliters of virus, each milliliter with more than a billion virus particles. His immediate supervisors and the USAMRIID safety officer refused to initiate a formal investigation, according to Levitt. He summarized his frustration:

> After three months of requesting administrative action and receiving none, my colleagues, including management level personnel, and I composed and signed a *Memorandum of Understanding* which documented a series of untoward events leading up to the mysterious disappearance of the chikungunya virus. Copies of the memorandum were sent to the Safety Officer and other administrative offices. To date, no satisfactory answer to how or why the virus disappeared or where it is, has ever been offered. . . .

[T]he missing virus material possibly could have impacted serious consequences on the surrounding environment.[19]

Levitt included a copy of the memorandum, dated December 30, 1981, with his statement to the Senate subcommittee. The document confirms what he claimed in his testimony. It was signed by Levitt and three colleagues and indicated that copies had been sent to his superiors and the institute's safety officer.

Colonel David Huxsoll, who had been commander of USAMRIID since 1983, disputed Levitt's testimony. He told the subcommittee that shortly after the incident "an informal investigation lead [sic] management to conclude that the material had probably been destroyed through autoclaving as was directed."[20]

Senator Levin then read into the record the conclusion of a 1986 inquiry by the army's inspector general: "The inquiry, therefore, substantiated Dr. Levitt's allegation that no investigation was conducted into the disappearance of the virus."[21]

The army included a description of the Chikungunya incident in its 1989 environmental impact statement as a response to a comment about the missing vials:

> This alleged incident involving missing virus occurred in 1981 and has been the subject of several intensive investigations and occasioned a visit to the laboratory by a member of Congress—all investigations and inquiries concluded that the vials were not lost, but had been destroyed inadvertently [sic], perhaps by the research team itself. Further, the allegedly misplaced vials contained an attenuated candidate vaccine virus and not a virulent organism.[22]

That is the army's response in its entirety. It suffers from error and innuendo. First, the missing virus incident was not "alleged" but was actual and denied by no one. Second, there is no evidence that "several intensive investigations" took place, as the army now claimed. Neither Levitt's nor the inspector general's contrary conclusions were mentioned in the response. Third, the suggestion that Levitt's research team destroyed the vials is gratuitous and without evidence. Finally, the army's implication that the virus was harmless—"attenuated" and "not virulent"—ignores the reason it was placed in storage in the first place: The weakened virus had proved more virulent than expected.

Whether or not someone removed the vials to rid the army of embarrassing evidence may never be learned. But the official explanation suggests, at the least, an effort to discourage serious inquiry.

LEVITT'S PERSPECTIVE

Since leaving USAMRIID in 1986, Levitt has remained in Frederick, Maryland, where he owns a delicatessen. In an interview in 1994, he expressed satisfaction with his work at the institute through the 1970s.[23] This changed with the reaction of his superiors to the Chikungunya project. When the experimental animals appeared to be dying from the control cells, he became more concerned about the cell line than the virus. Nevertheless, he "was getting a lot of pressure to drop this sort-of inquest about these cells, and just go on with the vaccine."

He refused. The fetal rhesus cells, some of which contained the virus, were placed in freezer storage in 1980. He and his assistant, Helen Ramsburg, went on to develop a successful vaccine with the use of a different cell line.

Although no longer working with the cells in storage, Levitt continued to tell his colleagues that he wanted to look into the reason the cells had killed the test animals. Then, "strange things started to happen."

At the beginning of 1981, he and Ramsburg retested some of the cells that had no virus in them. Now "all of a sudden they are loaded with Chikungunya virus." In July, a predawn fire broke out in the biological development suite at the institute. Early in September, a technician went to the freezer where some of the cells and virus were stored and found everything melted. The building engineer determined that someone had turned the thermostat from minus 70 degrees to higher than room temperature. That batch of vaccine was destroyed.

Two weeks later, on September 16, while taking an inventory, Levitt discovered that the remaining batches were missing from another freezer. "Every little vial, every large vial, every bottle were now gone." All these incidents were confirmed in the army inspector general's 1986 report, which had been requested by Senator Charles Mathias after Levitt had contacted the senator.[24]

Why would anyone want to remove the virus? Levitt's initial thoughts ranged from a break-in by terrorists to coincidence. But he believes the likeliest reason involved a cover-up by his superiors who had initially pressed him to use the fetal rhesus cells. The cells had been used in another army laboratory to make a vaccine against dengue fever. But that vaccine caused disease in the human subjects and almost killed one. "There was something wrong with those fetal rhesus monkey cells. I think some people before thought there was something wrong," Levitt said. The disappearance of the cells meant that they could no longer provide evidence against the administrators who insisted that they be used—or against USAMRIID itself.

Helen Ramsburg was one of the four people who signed the 1981 memorandum. She had worked with Levitt throughout the Chikungunya project. After retiring from USAMRIID in 1993, she offered impressions of the virus incident.

Ramsburg was astonished when the cells were reported missing. But she later accepted the institute's explanation that "in the process of cleaning out the freezers" someone must have autoclaved the cells (rendering them harmless) and discarded the remains.[25] The bigger problem, she said, was that this cell line was still being used: "That was what he was really concerned with, and I'm concerned with it too."

A Visit to Fort Detrick

Officials at Fort Detrick, not surprisingly, have a different view. During an afternoon in May 1994, I spoke with several there and asked about the Levitt case. Carol Linden, a civilian microbiologist, heads USAMRIID's Office of Research Plans and Programs. She said the official response to Levitt was made at the 1988 congressional hearing at which he testified. His allegations were investigated and "found to be untrue or resolved."

Colonel Arthur Anderson, a research physician at USAMRIID for 20 years, was somewhat more expansive. "It was Dr. Levitt's responsibility to properly handle those viruses," he said. "But he was absent when he should have been present."

Part of the reluctance by USAMRIID people to talk about the issue, Anderson said, was based on a legal settlement between the institute and Levitt when he departed. He was dismissed, but he then sued USAMRIID. The institute paid him a privately agreed-upon sum and agreed not to discuss certain sensitive matters about the case. Nevertheless, Anderson offered a psychological assessment about why Levitt sued the army:

> I think in the case of the reaction by Neil Levitt, he probably felt ashamed that he had not been responsible enough to supervise what he was doing. Rather than wallow in his shame, he converted the shame into blame. And the institute really was not prepared to accept the blame.

Is Anderson correct? Whatever the psychological issues, Levitt's version of the facts was largely documented. The incident did not help the army in its effort to market the safety of its BDRP.

THE GENERAL

Public perceptions (or misperceptions in the view of the military) seem to worry the army as much as any enemy's biological arsenal. "We need to do marketing," says Brigadier General Russ Zajtchuk, commander at Fort Detrick. Major Dale Vander Haam, whose job is to oversee the use of human research subjects, agrees: "We need to have you and other people ask questions. We want to put everything on the table that we can."[26]

How to allay suspicions. How to convince the public that activities at USAMRIID and elsewhere at Fort Detrick are safe and have nothing to do with offensive weapons development. How to make skeptics believe the army is no longer conducting secret tests on unsuspecting citizens. The response from these and other army officials was marketing; the professed technique, openness.

The general described his command responsibility as trying to prevent and treat disease among U.S. troops wherever they are deployed. In charge of the army's Medical, Research, Development, and Logistics Command, his job includes countering the effects of biological weapons. Is there a program to defend civilians as well as soldiers? "Obviously whatever we develop could be used by anybody," but real defense is a problem.

> In a chemical attack, at least you could have some forewarning because you have detectors that can give a reasonable amount of warning. As far as biological, if you suspect a threat you can encapsulate yourself in a particular suit which will hopefully protect. But it's not that easy to detect [biological agents]. If you get exposed to anthrax, by the time you detect it your symptoms may be so far advanced that the treatment becomes very difficult.

Why do army activities in the biological area seem to arouse more suspicions than do other army activities? Zajtchuk agreed with the premise but did not speculate on the reason. He turned instead to the revelations in the press a few months earlier about the 1950s-era radiation experiments on people. The press reports prompted him to inquire into the army's previous biological warfare tests. He was briefed on "everything that our people know," he said, and is confident that the work was "done on a volunteer basis with appropriate protocols."

What about the bacterial spraying over American cities in the 1950s and 1960s? Major Vander Haam mentioned *Serratia marcescens*, one of the bacteria used in the simulated attacks: "Scientifically, this is an organism that doesn't infect the human being." I mentioned that, to the contrary, many studies show that *Serratia marcescens* can infect people. Vander Haam then said he was not defending its use in past outdoor tests. Zajtchuk added that testing over populated areas in any case was no longer even contemplated.

Categorical as this seems, Zajtchuk demurred when I noted that open air testing was taking place at Dugway Proving Ground. When released, the army's bacteria and chemicals could drift beyond control. Zajtchuk: "I am speaking only about medical arenas. I mean I don't speak for the whole army." He was sure, however, that army officials at Dugway and elsewhere "are not hiding anything anymore, particularly if it involves human subjects."

USAMRIID

After the United States renounced its offensive biological program, several Fort Detrick buildings were turned over to the National Cancer Institute. Cancer research has been conducted there since the 1970s. At the same time, the base remains home to the army's principal institute for defensive biological research. Housed in a sprawling building, USAMRIID was completed in 1971 at a cost of $18 million. A 1994 brochure indicates that research at the U.S. Army Medical Research Institute of Infectious Diseases "concentrates on the development of vaccines, toxoids, and drugs designed to prevent casualties in the event of a biological warfare attack."[27]

Cheryl Parrott, author of the brochure, is USAMRIID's public affairs spokesperson. She insists that the work at the institute is open and spends most of her time responding to skeptical reporters, congressmen, and others. A two-hour meeting there with officials and research leaders underscored their commitment to improve the institute's image.

Why do people remain suspicious about biological research at USAMRIID? Because of past army tests? Because the work there is portrayed unfairly?

Vander Haam answered that people fear what they don't understand. His solution to the dilemma, by now the reigning mantra, is to educate the public—that is, marketing:

> We tend to understand bullets. We don't understand
> bugs and vapors. Our own soldiers in training are very
> fearful of biological weapons or chemical weapons. . . .
> So I think it's a matter of education.

Vander Haam expanded on the theme of public distrust. "I was at an educational curriculum fair in Pennsylvania and mentioned that I work at Fort Detrick. They said, 'Oh that's where they do the biological warfare studies.'" He seemed offended.

> I'm an American citizen even though I wear this uniform.
> But they say that we're the "others." Because you put this
> uniform on, or because you are a Department of the Army

civilian, you're one of the "others." But we still all live in the same community. We buy at the same store. You know, we're not the "others."

Vander Haam's comment accentuated what others at the institute had expressed in one way or another—that their work was misunderstood and underappreciated. Parrott asked if I thought it would help with their marketing effort to say "Yes, they did terrible things here in the past, and I'm glad we were not associated with them." I believed it would. Thoughtful observers had waited in vain over the years for an army spokesman to criticize past tests that endangered people.

Parrott's idea received little sufferance from others in the room. Colonel David Franz, then USAMRIID's deputy commander, had talked with the people who conducted testing over American cities, "and they were very proud of what they were doing." His empathy seemed more with the testers than the public exposed to the bacteria and chemicals. Vander Haam said that the standards of the time were different and he would find it hard to say that they did anything wrong.

Linden noted that even if individuals felt inclined to criticize, they could not speak for the army. Her own inclinations were reflected in her description of a recent television program. People were interviewed who had conducted above-ground nuclear testing in Nevada:

> They had one elderly guy, and I had to give him a lot of credit. He was not going to back down one iota. . . . He was absolutely convinced in his mind that whatever problems had been caused to people because of those tests, that was the cost of a higher good in the context of that time.

A Personal Experience

I do not question the sincerity of the Fort Detrick officials with whom I spoke. But the army's official response about past tests and current safety lapses seems self-defeating. The tradition of hyperbolic assurance that all is well—that damage to the environment and health is hardly conceivable—is demonstrably wrong.

Neil Levitt maintains that the army unjustly characterizes him as a disgruntled employee to cover its own failures. Because of an experience of my own, I cannot dismiss the possibility. It relates to a description that I published about building 470, known as "the Tower," at Fort Detrick.[28] The seven-story building housed large fermenter tanks and equipment to grow bacteria when the United States had an offensive program.

Writing in the mid-1980s, I indicated that the building had not been used since 1969, because it remained contaminated with anthrax spores. I compared the building to Gruinard Island in Scotland, which was off limits for more than 40 years because of a residue of anthrax spores from tests there in the 1940s.

I later learned that David Huxsoll, former colonel and USAMRIID commander, had castigated my description. On October 3, 1990, I wrote to Huxsoll, who had left the army to become a dean at the veterinary school at Louisiana State University. I explained that when I visited Fort Detrick in 1983 and 1984, more than a dozen people confirmed that the building remained contaminated. If building 470 had since been made usable, that was all to the good.

On October 12, I received a reply. Huxsoll wrote that he knew of no buildings at Fort Detrick that are off limits "because of contamination created by work in the offensive program, and I might add that I knew of none in 1984."

Three years after this letter, another Fort Detrick official contradicted Huxsoll's claim of safety. In a 1993 publication, the base's public affairs chief wrote:

> Efforts to demilitarize Fort Detrick included cleaning up Building 470, a facility more than seven stories tall, containing large tanks and ringed with catwalks from top to bottom. The primary agent grown at the pilot plant was anthrax, a dangerous organism which can lie dormant for thousands of years in a spore state.

> Anthrax spores can lodge in places such as cracks in concrete, and safety experts tried to decontaminate the building three times, he added.

After using special chemicals and monitoring devices . . .
the review team declared that Building 470 "appeared"
clear of dangerous organisms. They, however, noted that
because of the nature of anthrax it [*sic*] could not state that
Building 470 was 100 percent clean.

In 1988 the building was turned over to the Department of
Health and Human Services "and renovations were planned to
turn it into a useful support facility," the public affairs official
wrote.[29]

In January 1995, I made a telephone inquiry about the
disposition of building 470. My call was answered by Sheri
Hildebrand, secretary to the associate director of the Frederick
Cancer Center at Fort Detrick. She said that the building
remained empty. As far she knew, it may still contain anthrax
spores, and she was unaware of plans to renovate it for use. She
suggested that for more information I should write to Dr. Jerry
Rice, the associate director. I wrote to him on January 25 and
asked about plans, if any, for building 470. I never received a
reply.

Whatever the future of the building, the contention that it
is entirely safe is contradicted by the army's own public affairs
spokesman.[30] One can sympathize with the urge to market, repeatedly proclaimed by officials during my visit to Fort Detrick
in 1994. But untruthful marketing risks greater loss of trust.

The administrators of the biological defense program are
heirs to an institutional heritage of working in secret, sometimes
endangering unwitting citizens. The program may now incorpo-
rate more concern than in the past for the safety of the citizenry
and the environment. Nevertheless, breaches in the current
effort remain disquieting.

Many who work in the army program convey a sense of
sincerity and desire to do good. But they seem locked in an
institutional mold that precludes criticism of past or current
research activities. They will not acknowledge that several
experiments performed in the name of national security were
plainly wrong. It is this mind-set, not invidious behavior on the
part of any individual, that constitutes the larger danger associated with biological defense research.

CHAPTER 4

The Utah Experience

An inquiry in 1994 about activities concerning Dugway Proving Ground in Utah was answered with a small packet from the base's public affairs office. It indicates that the installation was established in 1942 and is "the Department of Defense central point for chemical and biological testing."[1]

Not a hint that work at the base has been a source of public unrest. In fact, political officials, scientists, and state residents have long worried that Dugway activities posed risks to people beyond its borders.

Dugway Proving Ground

The army has released bacteria and chemicals in more than a thousand open air tests in the years since World War II. Although testing over heavily populated areas was suspended after 1969, outdoor spraying with simulants has continued at Dugway. The army's presumption is that work there does not endanger anyone.

About 50 miles wide and 30 miles long, the facility is 70 miles southwest of Salt Lake City. Although Dugway is distant from large populations, the number of people living nearby has grown over the years. A few communities and Indian reservations are less than 20 miles away. Ranchers live closer and their livestock graze along Dugway's perimeter. Daily highway traffic in the Dugway vicinity exceeds 10,000 vehicles.[2]

The most dramatic consequence of testing at the facility occurred in 1968. Some 6000 sheep suddenly died in Skull Valley, 20 miles from the base. Scientists attributed the deaths to nerve gas released at Dugway, carried by unexpected winds.

Army records indicate that several ranchers in the area also became ill. They complained of flu symptoms, including nasal congestion, cramping, and diarrhea. Other army documents show that nerve gas can cause "runny nose and nasal congestion . . . nausea, vomiting . . . abdominal cramps, diarrhea." But army investigators concluded that the ranchers' symptoms could not "conceivably be related to organic phosphate (nerve agent) poisoning."[3]

Ray Peck, his wife, Connie, and their two children were among the people who became sick after the incident. As with others who complained to the officials, their symptoms were attributed to a viral infection. Twenty-five years later, a physician with the Centers for Disease Control suggested that the cause of illness could have been infection or nerve-gas exposure. "Unfortunately," he said, "there is no way to resolve this so long after the fact." Meanwhile, Peck and his family continue to suffer from headaches and numbness, problems they never had before the accident.[4]

Dugway authorities denied not only that the nerve gas caused the illnesses but that it killed any sheep. A year after the event, however, the army paid $1 million in damages to the sheep owners. But it still officially refuses to admit responsibility. Recent Pentagon references to the incident describe the sheep as "allegedly killed" by the nerve agent.[5]

Not all knowledgeable people are equivocal. A retired Chemical Corps officer who worked at Dugway in 1968 spoke on condition of anonymity. The sheep died from VX, a nerve agent the army released over the base, he told me in 1995. "We killed them, and we know we killed them."

The Myth of Safe Simulants

After the sheep died, the army stopped using highly toxic chemical or biological agents in open air testing at Dugway. But it continued spraying simulants. The official contention is that no one has ever been harmed by the army's tests with simulants, at Dugway or elsewhere.

Nevertheless, during 45 years of open air testing, the army stopped using certain simulants for reasons of safety. Their retirement from use was tacit acknowledgment that they could be causing disease and death. In fact, information about the danger of each simulant agent was available in the scientific literature long before the agent was proscribed.

This was the case in the 1950s when the testers ceased using the fungus *Aspergillus fumigatus* as a simulant. The fungus had been known for nearly a century to cause aspergillosis, a disease that can be fatal. Similarly, in the early 1970s the army stopped using zinc cadmium sulfide. Cadmium compounds had been recognized for years to be toxic to human organs and a potential cause of cancer, as noted in Chapter 2.

Later in the 1970s, the bacterium *Serratia marcescens,* a source of infections that can lead to death, was taken out of service as a simulant. In the 1980s, dimethyl methylphosphonate, a chemical known as DMMP, was removed from use in outdoor tests. Like the other biological and chemical simulants taken out of service, it was belatedly acknowledged to be toxic.

The suspension of these materials from the open air testing program raises several questions. Because the health of the people previously exposed was never monitored, no one knows whether they suffered ill effects. The army continues to spray other simulants outdoors at Dugway. It calls them harmless, just as it once described the now-forbidden agents, but will today's simulants also eventually be forbidden?

EARL DAVENPORT

The potential for problems from simulant testing is exemplified by the experience of Earl Davenport. He went to Dugway in

1958. As a 19-year-old army recruit, his job was to deliver supplies to toxic field test areas. After leaving military service in 1960, Davenport continued to work at the base intermittently for three decades. Over the years, donned in protective gear, he handled a variety of chemical and biological warfare agents. Although uncertain whether they caused him problems, he is convinced, ironically, that a simulant has.

In July 1984, Davenport took part in an outdoor test to see if a laser system could detect nerve agents. He was operating a sprayer to release a simulant of a nerve agent into the path of a laser beam. The simulant was DMMP. "During the test I noticed a sudden shift in the wind direction and quickly cut off the sprayer," he told a Senate committee in 1994, "but, before I could don my protective mask, a cloud of the chemical covered me."[6]

Davenport wiped his skin and left the test site to shower. A medic checked him and found nothing untoward. "I wasn't too concerned about getting hit with a simulant," he recalled. The agent was "practically nontoxic" according to a safety sheet the army provided, and he "trusted the army's assurance."[7]

The next day, Davenport began to feel ill. When his condition failed to improve, he was sent to the University of Utah Hospital in Salt Lake City. There he was diagnosed as having bronchial asthma probably caused by exposure to DMMP.[8]

His condition worsened over the years, and, after a heart attack in 1988, he was removed from work dealing with chemical and biological agents. In 1993, he stopped working altogether. Meanwhile, he obtained documents from the files at Dugway and concluded, according to his Senate testimony, that "the government had underestimated the health hazards of DMMP." Based on the documents he found, "underestimated" is an understatement.

THE DMMP PUZZLE

On April 6, 1984, the chief of the army's toxicology division assessed recent studies reporting toxic effects of DMMP on animals. This was *three months before* the test in which Davenport was exposed, and the army's toxicologist concluded:

The lack of a no-effect dose . . . makes it impossible to establish a safety factor for use of the material. Based on these considerations recommend disapproval of DMMP for simulant testing.[9]

Despite the recommendation, the army continued to use DMMP as a simulant in open air tests. In 1986, another report on the chemical's "toxicity status" was issued from Aberdeen Proving Ground, Maryland. Adverse findings in animal studies led to the following observation about DMMP:

Considering the positive mutagenicity findings, the severe, dose related male reproductive hazard and the current finding of some carcinogenicity and renal toxicity in the male rat, the development of an acceptable PEL [permissible exposure limit] seems highly unlikely.

The document recommended "that an alternative chemical agent simulant be developed to replace the DMMP currently used in military applications."[10]

Two years later, DMMP was still being sprayed outdoors. Another memo from another safety officer:

DMMP has been determined to be a mild carcinogen and potent renal toxin. Because of these hazards there is grave concern over the release of DMMP into the environment and exposure to liquid and vapors.[11]

In 1994, Dugway's public affairs office said: "Since 1988 the use of DMMP has been reduced significantly because the recommended exposure limit required protective clothing criteria and engineering controls that were so astringent [sic] that DMMP lost most of its value as a simulant."[12]

The chemical is no longer on the army's list of outdoor nerve-agent simulants. But DMMP's status in the testing program remains unclear. Despite the safety officer's expression of grave concerns about hazards from exposure, the army says only that its use has been "reduced significantly."

Whether or not Earl Davenport's illness was caused by DMMP cannot now be proved. But before he was exposed to the chemical in July 1984, he was documented as healthy. A routine physical examination at Dugway in January 1984 found his health "good," according to his medical report, and that he was using "no medications."[13]

Davenport's next routine medical report, in 1985, however, indicates problems with his lungs and chest and refers to correspondence from the pulmonologist. In subsequent annual reports, pulmonary problems are noted, including asthma and wheezing. After a myocardial infarction in 1988, he was seen by doctors for arrhythmias, chronic bronchitis, and emphysema and in 1993 was advised to stop work.

Believing his health had suffered because of exposure to DMMP, Davenport filed for workers' compensation. In July 1993, a Department of Labor claims examiner determined that the evidence "fails to establish that the claimed medical condition or disability is causally related to the injury."

In support of his decision, the examiner noted the conclusions of a physician to whom his office had sent Davenport. The physician determined that, because Davenport was a smoker, his smoking "history is sufficient to explain all of his current abnormalities."[14]

Perhaps Davenport's exposure to DMMP did not cause his health problems. Smoking may have been the exclusive cause, as the claims examiner and his designated doctor maintain. But perhaps not, and therein lies an ethical dilemma.

CARELESS POLICIES

The danger of the simulant DMMP was known by army testers from the time they started spraying it outdoors in the early 1980s. Despite a succession of memoranda noting its toxicity, the chemical was used extensively until 1988. (Although its use has been "reduced significantly," where and how it is now being used are not clear.) Moreover, Davenport's pulmonary illness began immediately after his exposure. Is it not fair to assume that the army's chemical played a part in his health breakdown? If so, the government seems obliged to provide at least partial compensation.

In a broader policy perspective, the public has reason for concern on two levels. First, the risks posed by the open air tests with simulants were often ignored in the past. It is unclear that they are fully appreciated today. A 1993 statement from Dugway's public affairs office implies that testing with simulants remains harmless: "No specific safety controls or protection are required for testing with simulants."[15]

Second, the army invariably denies that its simulants ever caused problems. This, even while removing several from service in tacit recognition of their risks. Can we expect a different army response in the future if people complain that simulants used in current tests have caused them illness?

If an outbreak of plague occurred in the Utah area, the Pentagon could be expected to deny responsibility. Yet a certain organism that was recently declared a simulant might draw suspicion.

Yersinia pestis is the organism that causes bubonic plague, a disease that has killed millions during the past thousand years. Known as the black death, a fourteenth-century pandemic wiped out three-fourths of the population of Europe and Asia. The bacteria are usually transmitted to human beings by fleas from infected rodents. Victims suffer high fever, chills, and delirium. Their lymph glands bulge out, filled with leaky pus. Then the skin becomes black from underlying hemorrhaging. Antibiotic treatment soon after infection will likely prevent death, but untreated victims commonly die in a few days.

Plague bacteria are considered a potential biological weapon. A 1991 news release from the Dugway Public Affairs Office announced that the organism would be used in a forthcoming indoor experiment. While calling the strain of *Yersinia* to be used (EV76) "relatively harmless," the army properly described it as pathogenic.[16] But two years later, the army shifted categories, declaring *Yersinia pestis* (EV76) to be a simulant.[17]

The EV76 label indicates a strain of *Yersinia* that has been deprived of iron. In this form, the bacteria are less threatening. Considered a simulant, they now might be used outdoors, for example, to assess the effectiveness of a detecting apparatus. (If the apparatus responded to the attenuated organisms, it would presumably do the same with the more virulent form.)

What the army does not acknowledge, however, is that if iron is introduced into the bacteria they may become highly pathogenic again. From a standard textbook on microbiology: "In *Yersinia,* the importance of iron in the degree of pathogenicity has been known for many years. . . . Avirulent nonpigmented organisms may be restored to their original expression of virulence for mice by providing an excess of free serum iron."[18]

Zell McGee, professor of medicine and pathology at the University of Utah Medical School, is appalled that attenuated *Yersinia pestis* is being used as a simulant. He is familiar with the mouse studies of avirulent strains. "But if the mice took in iron in drinking water, they got the effect of the disease." He believes that similar consequences could occur with people.[19]

The army's evident failure to account for this possibility is troubling, as is its redefining the test organism as a simulant. As previously noted, Dugway officials indicated that no special precautions are required when testing with simulants.

Another contradiction arises from a statement about emergencies. News releases commonly include this wording: "An emergency response plan is in place and training has been complete." Further: "Coordination of the emergency response plan with the Utah Department of Health and Department of Public Safety as well as the Tooele County Emergency Management Director has occurred." (Dugway Proving Ground is in Tooele County.)

Replying to an inquiry about the statement, Dr. Susan Mottice of the Utah Department of Health said there was an established procedure for emergencies at Dugway.[20] But Kari Sagers, the Tooele County Emergency Management director, answered differently. Her work focused on the disposal of chemical weapons at the Tooele post, 40 miles from Dugway, she said. When asked about coordinated emergency responses with Dugway she answered:

> We don't have joint drills or extensive exercise
> preparations. To be honest, I'm not real familiar with the
> biological mission or capabilities at Dugway. I'm sure they
> try to preclude anything from happening. But I'm not
> familiar with it.[21]

Apart from the effects of the army's tests on the general public, research using volunteer subjects also suggests reason for concern. A test of outerwear was conducted in 1993 on subjects who presumably had given their informed consent.

The 1993 Clothing Test

Standing naked outdoors is not likely to appeal to many people, even in a sparsely populated area. Besides, western Utah can be cold in November. So when 12 research subjects learned that their disrobing would be indoors, they were relieved.

As employees of the Lockheed company under contract at Dugway, the men's responsibilities included acting as "volunteers" for army experiments. The November 1993 test was to assess the ability of special outerwear to resist penetration by chemical agents. The subjects would be outfitted in battle regalia, perform exercises in the midst of chemical clouds, and then strip and submit to skin examination. Because the test included the spraying of chemicals outdoors, a formal environmental assessment was required. Completed in July 1993, the document was titled "Environmental Assessment for Battledress Overgarment Penetration Study, Phase 1—Field Test, at U.S. Army Dugway Proving Ground, Utah."[22]

After the test, Matthew Brown, a reporter for the *Deseret News* in Salt Lake City, sought to interview the test subjects. Dugway authorities arranged for him to meet with two in the presence of an army representative four months later.

Tests were conducted for 70-minute periods during four days, and the chemical cloud was so thick that "you could not see your hands." Dressed in outerwear and masks, the subjects had to engage in physical exercise, running in place and in figure-8 patterns.[23] Afterward they went into a trailer where an ultraviolet ("black") light was used to measure the concentration of chemicals on their clothing. They then disrobed. Sections of their clothing were cut and analyzed, and their skin was examined for traces of the chemicals. This was followed by baths and showers.

In the interview with Brown, the two subjects expressed confidence about their treatment during the test. But several questions remain unanswered about how the environmental assessment was applied, the nature of the test, and the experience reported by the subjects.

The subjects never saw the 63-page environmental assessment (EA). It begins by stating the purpose of the test and ends with a copy of the "volunteer consent form" the test subjects were to sign. The EA concluded that the test would have "no significant impact" on health or environment. The EA also noted that "each volunteer will be informed of potential safety and health hazards in test conduct." Moreover, "each volunteer will read and sign a volunteer consent form."

Yet, amazingly, no information about the chemicals to be sprayed at the subjects was on the consent form. Indeed, when Brown asked the two subjects if they had been told about the chemicals, they recalled hearing about them only *after* the test.

Perhaps the most important section of the EA concerning safety was the description of the four chemicals. Although not deemed highly toxic, each was capable of causing problems. Syloid can irritate the skin and mucous membranes. Tinopal can be corrosive to the eyes. Exposure to a third agent, uranine, prompted a manufacturer's advisory, which was included in the EA: "Remove contaminated clothing and shoes immediately. Wash affected area with soap or mild detergent and large amounts of water until no evidence of chemical remains (approximately 15–20 minutes). Get medical attention immediately."

The EA also included a manufacturer's warning about the fourth test substance, tetraethylene glycol:

Acute Effects: May be harmful by inhalation, ingestion, or skin absorption. Vapor or mist is irritating to the eyes, mucous membranes and upper respiratory tract. Causes skin irritation. Exposure can cause nausea, headache and vomiting.

First Aid: In case of contact, immediately flush eyes or skin with copious amounts of water for at least 15 minutes while removing contaminated clothing and shoes. If inhaled, remove to fresh air. If not breathing give artificial respiration. If breathing is difficult give oxygen.

The commentary on tetraethylene glycol makes a further provocative observation: "Ethylene oxide is used in tetraethylene glycol production and may be present as a contaminant. Ethylene oxide is an experimental carcinogen." Other than citing the quantity of ethylene oxide that causes cancer and death in experimental animals, nothing more is said about the chemical.

None of this was conveyed to the subjects before the test. Even if they had been given the information in advance, perhaps they would have participated anyway. In the interview with Brown, two of them said they would have. Nevertheless, providing risk information to test subjects after an experiment is to stand the notion of informed consent on its head.[24]

Citizen Unrest

The army's testing program at Dugway Proving Ground provoked unusual unrest that stemmed from a proposal in the 1980s. Controversy began in 1984 over the manner in which the Pentagon sought funds to build an aerosol test facility there. Rather than have the planned laboratory subjected to public scrutiny through congressional hearings, the army sought a reallocation of funds from existing projects. The facility was on a list of innocuous projects including military housing and garages.

THE HIGH-CONTAINMENT LABORATORY

Two months after the reprogramming had been approved by the ranking members of the House and Senate appropriations committees, one of them, Senator James Sasser, withdrew his assent. He belatedly recognized that the facility could be used "to test offensive biological and toxin weapons, a capability which is prohibited by a 1972 treaty."[25]

Sasser's concerns were echoed by others in Congress, the media, and the public. Responding to a suit by the Foundation on Economic Trends, the army issued an environmental assessment. In 1985, a federal court judge ruled that the assessment was superficial and enjoined the army from proceeding with the test facility until it provided a thorough environmental impact statement.[26] The army reluctantly agreed but indicated that the statement would require several years to prepare.

By the time the army issued a draft environmental impact statement on the test facility in January 1988, discontent among Utah residents had turned to resistance. Criticism focused on the army's plan to build the facility at the highest containment level possible, BL 4. (BL signifies "biosafety level," and the numeral 4 indicates that it would be the most stringent of four possible containment levels.) The designated degree of containment would allow for experiments with genetically engineered organisms and other pathogens for which no vaccine exists. The army said it had no plans to test such agents, but just wanted an extra measure of safety.

Many Utah residents remained unconvinced. Elected officials from both political parties voiced misgivings, including Republican Senator Orrin Hatch and Democratic Congressman Wayne Owens, who represented Salt Lake City. The state's Republican governor, Norman Bangerter, announced that he "adamantly opposed" the project.[27]

Salt Lake City's major newspapers applauded the governor's stance. Their editorials cited past army projects about which the public could feel little confidence: nuclear tests that may have caused cancer, the death of 6000 sheep from nerve gas that floated out of Dugway, and the simulated biological warfare attacks throughout the country.[28]

Protests continued through 1988, as hundreds of people attended meetings to express opposition to the laboratory. More than 140 University of Utah biologists and physicians signed a petition to protest the army's plans. At the end of the year, the army retreated. It withdrew plans for a BL 4 facility at Dugway and announced that it would build a lower-level BL 3 laboratory instead.

Critics of the facility remained uneasy. Cedric Davern, a biologist at the University of Utah, and Barbara Rosenberg, of the Memorial Sloan-Kettering Cancer Center, argued that even a BL 3 laboratory was unnecessary. They maintained that indoor defensive testing could be done with simulants that require less safety containment.[29] Nevertheless, public agitation began to taper off. But in 1991, when the army announced plans for further tests at Dugway, protests were revived.

New Testing

The concerns were prompted by an army news release on May 2, 1991. It announced that in June it would test "simulant materials" outdoors and pathogens indoors at Dugway. The purpose was to evaluate biological detector systems. The "simulant materials" included *Bacillus subtilis* and coliphage MS2 virus, which would be sprayed into the open air.

Other more infectious agents would be tested in containment areas: *Yersinia pestis,* which causes plague, and *Coxiella burnetii,* the cause of Q fever. Three toxins also would be tested indoors: botulinum toxin, staphylococcal enterotoxin B, and mycotoxin T-2.[30]

The army's announcement was widely covered in Utah's news media. The *Deseret News* reported the issue as one of risk versus gain, to be "mulled over by Utahns concerned about testing of deadly germs and toxins—and potentially hazardous simulants."[31] Although the proposed tests were reportedly approved by a biosafety group of federal and state officials, the state's Citizens Advisory Committee for Dugway Testing was not consulted.

The governor had appointed the committee in 1989 to work with Dugway officials on testing protocols. A member of the committee, Kenneth Buchi, a professor of medicine at the University of Utah, said the army seemed to have ignored this understanding.[32] University of Utah biologist Naomi Franklin doubted the tests could be of value, because a soldier would have to know what germs were in an enemy's arsenal before knowing what to test for.[33]

On July 1, 1991, a citizens group called the Downwinders filed suit in federal court to prevent the testing. The organization had led efforts to seek accountability for the health consequences of earlier nuclear tests. Now it claimed the planned biological tests showed a "wanton disregard for the health and safety of the citizens of the State." Downwinders director Preston

Truman said that state residents "will not stand idly by and serve as guinea pigs in yet another deadly experiment."[34]

A motion for injunction, prepared by former Utah Senator Frank Moss, noted the legal dilemma that victims of past nuclear and biological tests had faced. Courts have consistently held that the government is not liable for damages from its defense programs. Therefore, residents who might suffer from the effects of current tests "have no rights once they have been injured." The suit called on the court "to enforce their rights before they are injured, by the strict enforcement of health, safety, and environmental laws and regulations."[35]

The case was heard by District Court Judge Bruce S. Jenkins. He refused to grant an injunction, and in 1995 dismissed the suit. Jenkins conceded that organisms could escape from Dugway and harm people, but the possibility was "too speculative and hypothetical" to be an "injury in fact." Steven Erickson, a spokesman for the Downwinders, insisted that the judge's decision does not absolve Dugway of "responsibility for past, present, or future operations of their biological-weapons and/or defense-research programs."[36]

MEDICAL UNEASE

Among the lapses cited in the suit was the army's failure to train regional medical personnel to treat people infected with any of the agents. In June 1991, the University of Utah Medical Center in Salt Lake City informed army officials that it would not honor a contract to treat patients or perform autopsies on victims from Dugway who were contaminated as a result of the tests. The decision was taken because the army had not fulfilled its agreement to train hospital staff in the medical management of patients affected by chemical and biological agents.[37]

In response, the army agreed to begin a training program. The training amounted to two events. The first was a lecture that Kenneth Buchi characterized as "50 minutes of public relations and 10 minutes of medicine." Another physician in attendance, Zell McGee, said the university's physicians remained "completely unprepared" to deal with a biological testing accident at Dugway.[38]

The second event was a tour of Dugway in August for 22 physicians, nurses, and administrators from the University Hospital. Army personnel staged an accidental chemical munitions "leak" and treated a "victim." The visitors then received a technical briefing.[39]

Several doctors said afterward that they felt no better prepared to deal with victims of a testing accident. Nevertheless, Dugway officials issued a news release suggesting the army's obligation had been fulfilled: "An emergency response plan is in place and training is complete."[40]

The Citizens Advisory Committee; The Technical Review Committee

Although appointed by Governor Bangerter in 1989 as a result of public unrest, the Citizens Advisory Committee for Dugway Testing never received an explicit mission statement. The nine members at first decided to focus on health and environmental risks of biological defense testing. Several remained exasperated, however. They noted that meetings were held irregularly and the information they received was inadequate.

One member, Nancy Melling, who was also a member of the League of Women Voters, described the committee as "a rubber stamp. . . . When Dugway was going to do a test, we would get notice, and then they would just do it," she said. She felt frustrated: "We never had a chance to sit down in advance of a test to really understand what they were doing, what their goals were, what they were using, so we could make any decision."[41]

In 1993, the newly elected governor, Michael O. Leavitt, reconstituted the committee. Now called the Dugway Technical Review Committee, it would be an advisory body to the governor, not to the general public. The chair went to the state's newly appointed science adviser, Suzanne Winters.

By the end of the year, Winters felt there was more trust between the committee and the army. The committee had decided to meet every three months, and the army agreed to have members participate in on-site inspections. Unfortunately, she said in a conversation in 1995, the arrangements did not hold. When members tried to visit, the army "did not make it easy."[42]

Repeated cancellations of scheduled inspections left some members dispirited. "The committee has sort of come and gone for me," said Jay Jacobson, an infectious disease specialist and medical ethicist.[43]

The only member who has served since the original committee's inception is Susan Mottice, director of molecular biology for the Utah Department of Health. Also a member of Dugway's Safety Committee (which is unrelated to the governor's citizens/technical committee), she has visited the base many times. She believes that there is room for criticism about some of the laboratory work at Dugway but that fundamentally the facility is soundly administered.[44]

Mottice strongly supports the army's research program, especially in regard to detector systems for biological weapons. How could a detector identify genetically engineered organisms or other surprise agents? "There are only about 30 organisms that could be used as weapons," she believes. Moreover, genetically engineered organisms would have surface characteristics in common with related organisms. This presumably would make them detectable.

Mottice's rationalization for the program parallels the army's. Yet, even if a detection system for a wide variety of agents somehow were possible, she acknowledged that implemention for a large population would remain a "big problem."

In contrast to Mottice's comfort about Dugway's activities, Kenneth Buchi, a former member of the governor's committee, details his concerns. He first became interested when the army put out its 1988 draft environmental impact statement concerning the proposed BL 4 laboratory. As a member of the Environmental Health Committee of the Utah Medical Association, he was asked to review the statement.

Like others in the Medical Association, Buchi felt important questions were not being answered: "The concerns mostly revolved about doing BL 4–type work, and if they were not going to do it, why were they proposing to build a BL 4 lab? And if they were, they really did not address how they were going to prevent problems with BL 4 organisms." The army later withdrew its plans to build the highest containment laboratory, but Buchi remains skeptical about the army's intentions. "They have not built up any kind of history that would allow one to trust them."[45]

One theme the governor's committee tried to address before and after Buchi's departure is the army's reluctance to admit mistakes. Minutes of the September 1993 meeting, for example, refer to a Dugway report claiming that no "incidents," or accidents, happen there. One unnamed committee member expressed doubts because "most laboratories, in a normal working environment, have incidents." The committee's disturbing finding was that

> Some people at Dugway don't feel that they can bring up any safety issues without some reprisal. . . . [I]f they admit to an incident or a mistake they are concerned that they may be fired.[46]

Dugway Proving Ground is not the only U.S. location where biological defense testing is conducted. But the reaction of Utah residents underscores suspicions about the army's research program. It is emblematic of public anxiety that adheres to biological and chemical warfare issues.

The widespread opprobrium associated with these weapons invites skepticism about army activities in these areas, including those for defense. As this chapter and the preceding two demonstrate, suspicions have sometimes been warranted. Past experiments posed risks to millions of citizens. Today, military research is apparently conducted with greater caution, but questions about safety in the research program persist. Moreover, the army's reluctance to acknowledge mistakes, past or present, hardly engenders a sense of trust.

What to do? A frank admission by the army of ethical and safety lapses in past activities would be helpful. The 1975 inspector general's report citing the army's failure to adhere to informed-consent guidelines (discussed in Chapter 2) was a refreshing, if rare, acknowledgment. Such candor from within the institution can only enhance its credibility.

Also, the army's hyperbolic assurances about the safety and conduct of current research (Chapter 3) are self-defeating.

People are fallible, as everyone knows. Portraying the Biological Defense Research Program as virtually beyond reproach is bound to inspire skepticism.

Finally, today's BDRP is reputedly open and unclassified. Yet some people remain unconvinced. To help mollify concerns, the Pentagon should include appropriate outsiders in overseeing its research activities. This does not necessarily mean, as some suggest, placing all research in the hands of a nonmilitary agency, such as the National Institutes of Health. But inviting civilian scientists and ethicists—including skeptics—to participate in overseeing the program would be good for the military and the nation. Public assurances by credible outsiders would go far to allay mistrust. A national version of the Utah advisory committee—as it was *supposed* to work, with unfettered access to information—would be a good start.

Programs to defend the nation lose their moral force if they endanger and create anxiety among the citizenry. The challenge of matching moral behavior to national interest is taken up in the next part of the book as well. If defensive activities are worrisome, confronting the possibility of a battlefield attack with chemical or biological weapons can be terrifying. Iraq's war with Iran in the 1980s and the Persian Gulf War in 1990–91 are poignant reminders.

Lessons from Iraq

CHAPTER 5

IRAQ'S WAKE-UP CALL

esert Storm began January 17, 1991, with allied air attacks on Iraq. During the 43-day war, U.S.–led coalition forces used the most technologically advanced equipment available—cruise missiles, stealth aircraft, electronic jamming equipment, night-vision glasses. But one of the most insistent symbols of the war was a throwback of 70 years: the gas mask.

After invading Kuwait on August 2, 1990, Iraqi President Saddam Hussein rejected demands by the United States and other nations to withdraw. With UN Security Council endorsement, President George Bush began organizing a multinational military effort to end Iraq's occupation.[1] By the time the war ended on March 1, 37 countries had contributed to the coalition effort. The United States provided nearly 700,000 troops, and other countries 200,000, largely from Britain, France, and a few Arab states.[2] The troops anticipated being attacked by weapons that had only recently reemerged as a serious battle concern.

"Saddam Hussein has used chemical weapons both in war and on his own people," Prime Minister Margaret Thatcher reminded the British Parliament two months before the Gulf War began. "He also, we believe, has at his disposal biological weapons."[3] Not since World War I was there such publicly pronounced near certainty that Western military forces would face this kind of weaponry.

As they rehearsed for battle, coalition troops were seen in the media wearing gas masks and other protective equipment. But the accompanying narratives rarely suggested a sense of uniqueness about the image. Nor did they ask how the world had arrived at this state of affairs. The prospect of U.S. troops under chemical or biological attack now was anticipated with seeming resignation.

In condemning inadequate news coverage of the war, John R. MacArthur derided a *Time* magazine illustration, which he called the war's "signature emblem." It depicted, he said, an android soldier, a "cartoonish human form clothed in American desert camouflage and gas mask, striding forward with an M16A2 rifle."[4] But for MacArthur, as for the media he criticized, wearing the gas mask merited no special mention. It simply had become part of the contemporary military image.[5]

George Bush also was connected to the new image. *New York Times* reporter Maureen Dowd wrote of nervous anticipation in Washington as the nation entered the war. The day the war began she noted that "the President was accompanied by a Secret Service agent who held a gas mask in a green military bag."[6] Why Bush, 7000 miles from the battlefield, was deemed vulnerable to chemical or biological attack is not discussed in the article. Dowd called the president's gas mask "the day's most ominous image," symbolizing that war had begun. Ominous it was. But not simply as Dowd's metaphor for America's entry into the war.

The broader implications of the mask evidently escaped the reporter's notice. Dowd made no other allusions to chemical or biological warfare in her article. The absence of such discussion represented the disturbing prevalent attitude. The mask no longer seemed extraordinary. (Five years later, the mask appeared just as unremarkable. A 1996 news story about tensions between North and South Korea began with a description of a U.S. lieutenant with his "olive-colored gas mask ready for instant use beside him."[7])

In fact, the president's mask symbolized a dual policy failure on the part of the United States and the international community. Companies from Western countries were instrumental in helping Iraq develop chemical and biological weapons programs. Meanwhile, in the 1980s, the nations of the world watched as Iraq used chemical arms against Iran.

Western Companies and Iraq's Weapons

A few Iraqi scientists could doubtless have begun rudimentary chemical and biological weapons programs without outside help. But the size of the Iraqi capability in the 1980s, and the speed with which it was developed, were possible only because of assistance from the West.

THE CHEMICAL PROGRAM

Iraq's interest in chemical weapons began in the 1970s. In trying to acquire a production capacity, it sought corrosion-resistant materials in 1975 from the Pfaudler Company in Rochester, New York. The purchase was ostensibly for a pesticide production facility.

Pfaudler questioned the massive size and lack of safety concerns in the Iraqi plans. When the company refused to meet the specifications, Iraq turned to ICI (Imperial Chemical Industries) in the United Kingdom. Recognizing that Iraq was seeking export-controlled chemicals that could be used to make gas weapons, Imperial also abandoned negotiations.[8]

By the late 1970s, however, Iraq had succeeded in obtaining components from firms in Switzerland, the Netherlands, Belgium, Italy, and, notably, West Germany. The German company Karl Kolb contracted to build the "pesticide" plant. Other German firms made a range of contributions—technical help from the Fritz Werner Company; pumps and chemicals from Water Engineering Trading GmbH; and vessels, centrifuges, and pipes from Quast GmbH.[9]

Saddam Hussein's hopes for a nuclear capability were derailed when the Israelis destroyed his Osirak reactor in 1981. In the aftermath, he accelerated development of his chemical weapons production center near Samarra, 45 miles northwest of

Baghdad. The installation was called the State Establishment for the Production of Pesticides (SEPP).

"The companies supplying SEPP formed a Western European telephone book," wrote Herbert Krosney, author of a study on the arming of Iraq. By the early 1980s, firms from France, Great Britain, Australia, and the United States were doing business there, as were companies from Poland, India, and other non-Western states. "The whole world was there," observed an engineer from Germany's Hammer Company in 1984 when he started work at SEPP.[10]

The companies provided experts and building materials, as well as equipment and ingredients to make the weaponry. Some 40 German technicians alone worked on the design and construction of the buildings. German court records and UN reports indicate that in 1982, facilities at SEPP were producing mustard gas and the nerve agents sarin and tabun. Still, Kolb company officials insisted that their work involved only a pesticide plant.[11]

The Hammer Company representative in Iraq, Bernd Meyer, indicated that in 1983 and 1984, the facilities were dramatically enlarged. In the installations that his company helped build, he saw bins containing chemicals sent from Dutch and other companies. The bins, according to their labels, included thionyl chloride and thiodiglycol, precursors in the production of mustard gas.[12]

Foreign technicians continued to maintain that they were helping with insecticide production. In 1984, when evidence irrefutably showed that Iraq was attacking Iran with chemicals, a few companies curbed their sales. Most showed no such scruples.

A 1990 study listed 207 firms in 21 countries that provided supplies for Iraq's unconventional weapons programs. Transactions continued through the Iran-Iraq war and after. Eighty-six companies were in the Federal Republic of Germany, the most in a single country. Eighteen were in the United States, and 18 in Great Britain. Seventeen firms were in Austria, 16 in France, 12 in Italy, and 11 in Switzerland.[13] American companies that were listed included Hewlett Packard, Alcolac International, and the Al Haddad Trading Company.[14] Some $2.8 billion in U.S.

grants apparently was passed through Italy's National Labor Bank to help finance Iraq's military buildup.[15]

Iraq's success in obtaining materials from foreign suppliers taught a dark lesson. Libya's chemical procurers watched closely. In consequence, political scientist Thomas Wiegele noted: "Perhaps the strongest demonstration to the Libyans was the ease with which [Iraqi arms acquisitions] took place. Not only was there no apparent attempt to stop the Iraqis, but arms merchants eagerly offered their wares for sale."[16]

Thus, the United States and others not only helped develop the Iraqi chemical and biological weapons programs, but they inspired countries to follow the Iraqi model. Controls by Western governments over such exports were feeble or nonexistent.

West German officials began to show interest in restraining German companies from involvement in chemical weaponry in the mid-1980s. The government issued regulations to require an export license for dual-use chemicals—that is, those that can be used for both civilian and military purposes. In November 1987, it even searched companies suspected of violating the regulations. In 1988 and 1990, however, German courts refused to uphold the government's charges that the companies acted illegally.[17]

The German parliament enacted legislation in 1990 prohibiting anyone from assisting in the development of chemical, biological, or nuclear weapons inside or outside the country. Opposition politicians, however, believed the law could not be effectively enforced.[18]

Meanwhile, several industrial countries had been meeting periodically since 1985 to encourage export controls of certain chemicals. By 1990, the Australia Group (named after the location of the first gathering) comprised 20 members, mostly from Europe. The group had devised a list of 50 industrial chemicals that could also be used to produce chemical weapons. In 1991, the group agreed that members should exercise full export controls over the chemicals; the next year, it agreed to consider controls on biological agents as well.[19]

As for the United States, in 1989 the Commerce Department strengthened export controls on chemical and biological agents with military potential. Twenty-three chemicals and five biological agents were to be restricted. The new controls were aimed specifically at Iran, Iraq, Libya, and Syria.[20] Although welcome, the restrictions, like those of the Australia Group, had doubtful effect on Iraq, which by then had its own production capability.

THE BIOLOGICAL PROGRAM

Before 1991, more attention had been given to Iraq's chemical weapons than to its nuclear or biological programs. This was largely because Saddam had demonstrated a formidable chemical capability against Iran. The magnitude of Iraq's nuclear program was not clear until after the Gulf War, when discovery of its advanced state surprised many experts.[21]

The extent of Iraq's biological weapons program also was uncertain before the Gulf War. In fact three years later, UN inspectors were still saying they knew less about Iraq's biological than its chemical or nuclear programs. The inspectors complained that Iraq was violating Security Council resolutions by not providing information in the biological area.[22]

The earliest public reports about an Iraqi biological warfare program appeared in 1988, toward the end of the Iran-Iraq war. United States officials confirmed news stories that Iraq was developing weapons that can spread typhoid, cholera, and anthrax. The work was being done at a facility in Salman Pak, about 40 miles southeast of Baghdad.[23]

The following year, news articles again noted that Iraq was working on biological warfare agents, adding that the U.S. Centers for Disease Control and Prevention (CDC) had sent virulent materials to Iraq during the 1980s.[24] Whether Iraq had yet stockpiled biological weapons remained unclear.

A review of exports from the United States that could have helped Iraq's biological program was included in a 1994 report by the Senate Committee on Banking, Housing and Urban Affairs. The report derived from the committee's authority over export administration. Its staff initially requested information from the Department of Commerce concerning biological materials sent to Iraq in the years before the 1991 Gulf War.

The staff then contacted a principal U.S. supplier, American Type Culture Collection (ATCC). Located in Rockville, Maryland, ATCC is a private, nonprofit institute that collects, preserves, and distributes cultures of microorganisms and other biological materials. Its board of directors includes representatives of 22 North American biological associations.[25]

ATCC gave the Senate committee a list of biological materials it had exported to Iraq between 1985 and 1989. Included were pathogenic bacteria, toxins, and other agents. The receiving agencies were Iraq's Atomic Energy Commission, Ministry of Trade, Ministry of Higher Education, and the State Company for Drug Industries. On May 2, 1986, for example, ATCC shipped 24 batches of agents, including two strains of *Clostridium tetani,* three of *Bacillus anthracis,* and five of *Clostridium botulinum.* The causes of tetanus, anthrax, and botulism, they have long been considered potential biological warfare agents.[26]

In 1994, ATCC's associate science director, Richard Roblin, told reporters that "We don't make decisions about which cultures are shipped" to Iraq. "We rely on the Department of Commerce. It's their job."[27] A year later, the institute's public affairs spokesman, Patrick Burke, amplified on the institute's work. Most of ATCC's specimens are used for basic research, he said, although "a lot are used as quality controls when testing patient samples, for medical purposes."[28]

Burke acknowledged that several items are hazardous and could be used for nefarious purposes, though many could no longer be sent to certain countries. With resignation he added, however, that biological agents are found in nature, and "if anyone really wanted to, they could get them from another source."

In May 1994, the CDC director provided Senator Donald Riegle a list of agents it had sent to Iraq from 1984 through 1989. Shipments included botulinum toxoid (listed as noninfectious) and the West Nile fever virus, a potential biological warfare agent.[29]

In July 1995, in response to a Freedom of Information Act request, the CDC informed me that the information it had supplied to Riegle was incomplete. It had sent Iraq more than 80 agents and associated materials not on the earlier list. They included *Yersinia pestis* (listed as avirulent), dengue virus, and

West Nile antigen and antibody. Thus not only were private firms sending potential biological warfare materials to Iraq, but so was a U.S. government agency. The policy was to send almost any country whatever it wanted.

Senator Riegle, chairman of the Banking, Housing and Urban Affairs Committee, criticized the Commerce Department for allowing the exports. He was especially vexed because a Pentagon report indicated that Iraq was developing biological weapons during the 1980s: "I think the U.S. Government approving exports of these materials to a government like that and to someone like Saddam Hussein violates every standard of logic and common sense."[30] The senator wondered whether biological and chemical materials from the United States later caused illnesses to American troops during the Gulf War, a subject discussed in Chapter 7.

The private companies can retrospectively be accused of unprincipled economic greed. But their eager cooperation would have been dampened had their governments objected. Before 1984, the United States, Germany, the United Kingdom, France, and others imposed no meaningful restrictions on exports to Iraq. Afterward, some countries were more attentive. But by the time most became serious about monitoring overseas sales, Iraq was producing its own weapons.

Failure of the West to curb exports to Iraq was only part of the problem. More ominous was the global failure to condemn Iraq's actual use of chemical weapons.

The Iran-Iraq War

In February 1979, Ayatollah Ruhollah Khomeini assumed power in Iran. The move followed the overthrow of Shah Mohammed Reza Pahlevi, who had ruled since 1941. The Shah's pro-Western repressive regime was replaced by the Ayatollah's anti-Western repressive regime.

Khomeini and his fellow clerics promised to impose their Islamic fundamentalism in Iran and beyond. To make matters worse for the United States, in November 1979, Islamic militants seized the U.S. embassy in Tehran. They held a group of Americans there, making a variety of demands for their release. After 13 months of captivity, 52 hostages were freed as the United States agreed to release Iranian funds in American banks. By then, attitudes in the United States had become thoroughly disdainful of Iran's new regime.

Under the Shah, Iran had depended on the United States for military supplies and training. With the establishment of the Ayatollah's anti-American regime, U.S. support was suspended. Saddam Hussein decided to capitalize on the confusion among Iran's new rulers. In September 1980, Iraqi forces occupied an area of Iran near the Persian Gulf. By moving into Iran, Saddam abrogated a 1975 agreement that ostensibly had settled a border dispute between the two countries. Saddam's move infuriated Iranian leaders, inducing them to set aside their own squabbles.[31] War had begun.

During the next eight years, Iraq and Iran alternately gained and lost the military edge in a cycle of battles. Many countries, including the United States, were not unhappy to see the war continue. Khomeini's energies were diverted from exporting his brand of fundamentalism. While overtly tilting toward Iraq, however, U.S. officials also secretly supplied weapons to Iran. News about the sale of antitank missiles and spare parts to Iran surfaced in 1986 in reports from a Lebanese-based magazine. The story adumbrated further revelations that became known as Irangate and Iran-Contra.[32]

But if observers professed shock to learn about the U.S. arms deal, a scandal with larger implications was taking place in full view, in which nations throughout the world were accessories.

IRAQ'S USE OF CHEMICAL WEAPONS

Both Iraq and Iran were parties to the 1925 Geneva Protocol, which prohibited the use of chemical or bacteriological agents in war.[33] Nevertheless, Iraq reportedly used chemical weapons on a limited basis in 1982.[34] In November 1983, it dropped mustard-gas bombs to try to dislodge Iranian troops who had occupied Iraqi territory the month before. The effort failed at first. After Iran captured Iraq's Majnoon Islands near Basra in February 1984, the Iraqis intensified their chemical attacks along with conventional arms. By March, Iran claimed that Iraqi mustard and nerve weapons had killed 1200 and wounded 5000 of its troops.[35]

United States authorities reported in March 1984 that Iraq was using chemical weapons. The State Department initially made the claim, and intelligence officials then said they had incontrovertible evidence that Iraq was using nerve agents against Iran.[36]

At the same time, UN Secretary-General Perez de Cuellar assigned a team of experts to investigate, the first of many UN teams to confirm that Iraq used chemicals. But neither the United States, the United Nations collectively, nor any other body of nations seriously protested. Indeed, two months later, ignoring Iraq's transgressions, the Security Council passed an Arab-sponsored resolution to condemn Iran for attacking commercial shipping in the Persian Gulf. The United States voted with the 13–0 majority.[37]

In November 1984, the United States reestablished full diplomatic relations with Iraq, which Iraq had broken after the Arab-Israel war in 1967. The pattern of behavior by the United States and other nations continued through the end of the war: reluctant acknowledgment of Iraqi chemical attacks; dealing with Iraq as though they never occurred.

In March 1986, the UN Security Council criticized Iraq by name for resorting to chemical weapons. The council's action was in response to the previous month's finding by UN experts that the Iraqis had used chemical weapons on many occasions. But the council's willingness to cite Iraq was unusual. It was the first and last time during the war that Iraq was accused by name in any UN resolution on the subject. Even then, the Security Council's disapproval seemed incidental to its undirected condemnation of "the prolongation of the conflict which continues to take a heavy toll of human lives."[38]

On July 7, 1987, the Security Council unanimously passed Resolution 598 calling for a cease-fire between the combatants. As had become customary, it failed to condemn Iraq's use of chemical weapons. Although Iraq agreed to the cease-fire as urged by the resolution, a bitter Iran demurred.

During the next year, Iraq stepped up its chemical attacks. Blistered skin and burnt lungs from mustard gas; choking, frothing, and agonizing death from nerve agents—all had become commonplace on the battlefield.

Iraq's liberal use of chemical weapons, according to Dilip Hiro, "had considerably dampened the [Iranian] volunteers' usual enthusiasm for combat."[39] Iranian authorities acknowledged that Iraq's aggressive use of chemicals was forcing Iranian troops to retreat. Robin Wright refers to an alleged incident in which Iraq used only smoke bombs, but "the psychological impact of chemical weapons was such that the Iranians had fled at the mere sight of the clouds."[40]

The psychological effects were at least as significant as the physical. Iranian troops became terrorized, and in July 1988, Iran agreed to a cease-fire under the terms of Resolution 598. John Bulloch and Harvey Morris describe the importance of chemical weapons in the decision.

> It seemed that Iraq's use of poison gas had a decisive effect
> on Iranian morale, and because of the various Western
> embargoes on the export of war material to Iran, the
> country was unable to equip itself with adequate supplies of
> chemical warfare clothing. Use of the protective suits was,
> in any event, impractical in the extreme heat of the
> southern front, so that when faced with a chemical attack,
> the Iranian forces had little option but to cut and run. It
> was as if the fervent Revolutionary Guards, who had so long
> proclaimed their readiness to die for Islam, had lost their
> will to fight, and slowly the Iranian war machine ground to
> a halt. . . .[41]

Shortly before the cease-fire, an Iraqi official acknowledged for the first time that his country had used chemical weapons. At a news conference in July 1988, Tariq Aziz, the foreign minister, referred to allegations that both Iran and Iraq had used chemicals. He claimed that Iran had used them first and that the "Iranians were the invaders of Iraq." No credible authority believed either contention. But even in the face of Aziz's historical distortions, his admission was unmistakable: "Sometimes such weapons were used in the bloody war, by both sides. It was a very complicated, bloody conflict. It has to be judged within the circumstances and the facts."[42]

Iraq's accelerated use of chemical weapons in the final year of the war was officially ignored by the UN Security Council. As in five previous UN-sponsored inquiries, investigators in 1988 confirmed that Iraq was using such weapons. Their report added that the use of chemical weapons by Iraq "has become more intense and frequent."[43]

Nevertheless, Sir Crispin Tickell, Britain's UN ambassador, said the Security Council "didn't want to upset the applecart" by criticizing Iraq before peace talks.[44] The United Nations' unwillingness to be specific seemed to bother no one. Press reaction was largely passive, as exemplified in a *New York Times* article, headlined "Chemical Arms Condemned." Dated August 26, 1988, the article was two sentences long:

> The Security Council today unanimously adopted a
> resolution condemning "the use of chemical weapons in
> the conflict between Iran and Iraq." The resolution did not
> condemn Iraq by name.[45]

That was the whole story. No inquiries about why Iraq went unnamed, no assessment of that fact's significance or its implications for the future.

Unlike many accounts of the Iran-Iraq war, that of Bulloch and Morris recognized the consequences of the world's silence about Iraqi chemical activities. Writing in 1988 after the cease-fire, they observed: "One of the most dreadful facts to emerge from the eight years of war was that chemical weapons could now be used in local conflicts. . . . The world awoke too late to what was going on."[46]

The message was not lost on countries that had begun to develop chemical and biological weapons. The dozen or so that had chemical programs in the late 1970s had grown to more than 20 a decade later. And despite the 1972 Biological Weapons Convention, which prohibited the possession of biological weapons, at least ten countries had biological programs by 1989.[47]

Military planners had witnessed the apparent effectiveness of Iraq's chemical attacks and the erosion of moral strictures against them. A telling observation was made by Iran's Rafsanjani after the Iran-Iraq war:

> With regard to chemical, bacteriological and radiological weapons training, it was made very clear during the war that these weapons are very decisive. It was also made clear that the moral teachings of the world are not very effective when war reaches a serious stage and the world does not respect its own resolutions and closes its eyes to the violations and all the aggressions that are committed in the battlefield.[48]

DID IRAN USE CHEMICALS?

Allegations were made later in the war, as noted in Aziz's statement, that Iran employed chemical weapons. But the evidence was equivocal. In 1988, UN investigators determined that chemical grenades ostensibly captured from Iran could have been Iraqi ammunition.[49] Journalist Martin Yant contended that President Bush had exaggerated Iraq's chemical threat because he was obsessed by Saddam Hussein. Yant's antipathies led him to fanciful conclusions. Ignoring evidence from UN investigators, he not only proposed that the Kurdish population in Iraq had been gassed by Iran, but questioned whether Iraq had chemical weapons "in any numbers at all."[50]

Yant was not alone in his selective reference to evidence about Iraq's chemical weapons, especially his doubts that they were used against the Kurds. About the claim that Kurds were gassed in Halabja, Edward Said, who teaches English literature at Columbia University, wrote: "At best, this is uncertain." Like Yant, he cited a U.S. Army War College study that suggested the Iranians may have been responsible.[51] One of the authors of the war college study, Stephen Pelletiere, a former intelligence officer, reiterated in 1992 that he was "fairly certain that Iranian gas killed the Kurds."[52]

Claims that Iran used chemical weapons in Halabja are a minority view. Reports by U.S. investigators that the Iraqis were responsible are supported by almost all analysts.[53] Perhaps the most insistent account appears in a report by Middle East Watch, a human rights organization. After sifting through captured Iraqi documents, examining forensic evidence, and interviewing hundreds of Kurds, investigators from the organization say that the evidence for Iraq's sole culpability is "crystal clear." Their report criticizes "writers in the United States [who] continue mystifyingly to insist" otherwise.[54]

Drawing on several news sources, Dilip Hiro cites Ayatollah Khomeini as saying in early 1988 that he would not use chemical weapons because "Islam prohibits its fighters from polluting the atmosphere even in the course of a jihad, holy war."[55] If true, this is an extraordinary example of moral restraint even while Iran was losing on the battlefield. In any case, the Ayotollah's morality at best was selective, for he had no compunctions about brutalizing opponents by other means.

Moreover, Iran was known to be developing chemical weapons at the end of the war. Anthony Cordesman, a military affairs specialist, wrote that Iran began a chemical weapons program in response to Iraq's chemical attacks. By March 1988, two Iranian facilities were producing chemical weapons.[56] If the Ayatollah was reluctant to use such weapons, others in the regime were less inhibited.

Akbar Hashemi Rafsanjani, who succeeded Khomeini after his death in 1989, said the war showed that weapons of mass destruction could be used with impunity. Addressing Iranian soldiers in October 1988, he said ominously that "we should fully equip ourselves both in the offensive and defensive use of chemical, bacteriological and radiological weapons."[57] Thus, irrespective of Iranian behavior during the war, the country later seemed bent on developing the full range of unconventional weapons.

DID IRAQ USE BIOLOGICALS?

If allegations that Iran used chemicals against Iraq are questionable, evidence that Iraq used biological agents is even more tenuous. Nonetheless, award-winning journalists William Burrows and Robert Windrem think the claim credible. They list diseases and symptoms that Kurdish physicians and Iranian officials found among victims of alleged Iraqi attacks: botulism, gas gangrene, anthrax, tularemia, very high fever, and bloody diarrhea. They also cite a 1987 Iraqi army document obtained by Kurdish guerrillas after the Gulf War that instructed forces to take an annual inventory of all "biological and chemical materials."[58]

The Iraqi document is far from clear, however, about the use of the inventoried materials. Moreover, the illnesses and symptoms demonstrate a universal problem about trying to determine if biological weapons have been used. The illnesses can come from natural sources or from infections from wounds or other injuries.

Mixed Signals from the United States

Whatever the uncertainties about Iraq's alleged use of biologicals or Iran's of chemicals, some matters are beyond dispute: Iraq used chemical arms years before Iran had a chemical capacity; it used them increasingly during the war; most of the world community watched in silence.

At the war's end, columnist Flora Lewis was among the few observers who took note of the world's nonresponse. She decried the "deafening silence of governments on Iraq's use of chemical weapons." The silence, she properly observed, encouraged other countries to develop chemical weapons because "the complicity of the world community with Iraq shows that can be done with impunity."[59]

In early September, the U.S. Senate passed legislation that would impose sanctions on Iraq for its attacks against the Kurds. But on September 22, when a House committee approved similar legislation, the Reagan administration announced its opposition. A state department official said the United States was engaged in discussions with Iraq, "which last week declared its opposition to use of chemical weapons."[60] The administration seemed willing to accept Iraq's word that it opposed the use of chemical arms despite six years of evidence to the contrary.

Bitter that the sanctions effort was killed, Democratic Senator Claiborne Pell blamed "special interests" for influencing the administration." In a speech on October 21, 1988, he said:

Agriculture interests objected to the suspension of taxpayer subsidies for agricultural exports to Iraq; the oil industry protested the oil boycott—although alternative supplies are readily available. Even a chemical company called to inquire how its products might be impacted.[61]

In 1989, the dangerous consequence of the years of silence began to receive more public acknowledgment. William Webster, director of Central Intelligence, worried that the Iran-Iraq war had ended "restraints" about the use of chemical weapons. He was concerned "that the moral barrier to biological warfare has been breached" as well.[62] Senator William Roth wondered how to recapture the power of "moral persuasion" in this area. And Senator Joseph Lieberman worried about "crossing another moral threshold, quietly losing our capacity for outrage and action."[63]

The importance of the moral component as a restraint against biological and chemical warfare was at last being recognized in official Washington. The next year, the United States and others would have to face in military terms the consequence of this lowered moral barrier.

Meanwhile, the Reagan administration's reluctance to punish Iraq carried into the new government under George Bush. Bolstering Saddam Hussein, not punishing him, would serve U.S. commercial and national interests. In October 1989, President Bush signed a secret directive to encourage private American companies to do business with Iraq. The directive included opposition to "any illegal use of chemical and/or biological weapons." But its overriding message was: "We should pursue, and seek to facilitate, opportunities for U.S. firms to participate in the reconstruction of the Iraqi economy."[64]

What Bush meant by "illegal use" was unclear. Iraq was a party to the 1925 Geneva Protocol that forbids use of these weapons in war.[65] It had signed, although not ratified, the 1972 Biological Weapons Convention that prohibits development or possession of biological weapons.[66] By the standards of these treaties, Iraqi actions were already illegal.

The directive was characterized by Sam Gejdenson, a Democratic congressman, as exemplifying "the administration's sole desire and policy to aid and abet Saddam Hussein."[67] The following month, in November 1989, the Bush administration extended $500 million in credit guarantees to Baghdad for commodities purchases.[68]

Naive thinking about Iraq's intentions reached beyond the government. A monumental example appeared in a column by journalist Milton Viorst. Diplomacy, not sanctions, would be sufficient to persuade Iraq to forgo using chemical weapons, he argued, apparently believing the unbelievable:

> The secretary of state and Congress are certainly right in seeking to halt the spread of gas warfare. But Iraq, having put down the Kurdish rebellion, currently has no war on its agenda, and it has pledged to abide by the Geneva convention in the future.[69]

Twenty-one months later, Iraq invaded Kuwait.

The Gulf War and the Chemical and Biological Specter

Saddam Hussein's invasion of Kuwait on August 2, 1990, surprised everyone. The move angered many Arab states, who viewed the act as a breach of Arab unity. The sense of betrayal was heightened because of past Arab support for Iraq, including refusal to acknowledge that Iraq ever used chemical weapons. Sheik Ali al-Khalifa al-Sabah, Kuwait's finance minister, later told an American journalist:

> We didn't want Iraq to collapse, and our sense of brotherhood drew us. . . . We said that Iran started the war. Hell, we knew who started the war. But we tried to cover every mistake the Iraqis made.[70]

Opponents of Iraq's move into Kuwait were primed by more than brotherhood or selfless opposition to aggression. The Gulf states, especially Saudi Arabia, now worried about their own vulnerability. The United States and Europe were unhappy that Saddam would control 20 percent of the world's oil supplies.

Four days after the Iraqi invasion, the UN Security Council enacted a resolution ordering a trade embargo against Iraq. The resolution was the first of twelve during the next eight months aimed at forcing an Iraqi withdrawal. The world seemed bent on reversing Saddam's action. Bush was infuriated by the occupation and likened the Iraqi leader to Hitler. (Elaine Sciolino observed that calling Saddam a modern-day Hitler may have enhanced his standing at home. During World War II, Iraqis applauded Germany's attacks on Britain and the Jews. "As Iraqis saw it, Hitler's only fault was that he lost the war."[71])

But the Bush administration's epiphany about the dangers of Iraq's chemical and biological arsenal was slow in coming. The month after Iraq occupied Kuwait, Secretary of State James Baker was still equivocal about the ultimate disposition of Iraq's unconventional weapons. In September 1990, Congressman Steven Solarz asked him what would happen to these weapons if Iraq evacuated Kuwait. Baker responded that the matter would have to be addressed as part of a new "security structure" in the region. A frustrated Solarz wondered rhetorically: "Is it possible to eliminate the Iraqi nuclear, chemical, and biological weapons programs without physically destroying them?"[72]

The administration seemed uncertain until the eve of hostilities. A last-minute effort to forestall military action against Iraq still seemed to offer a way that Saddam could keep his chemical and biological arsenals. When Baker met with Iraq's foreign minister, Tariq Aziz, in Geneva on January 9, he implied that as long as Iraq did not *use* these weapons against American troops, it could keep them.

> [I]f the conflict starts, God forbid, and chemical or
> biological weapons are used against our forces,
> the American people would demand revenge,
> and we have the means to implement this. This
> is not a threat, but a pledge that if there is any
> use of such weapons, our objective would not
> be only the liberation of Kuwait, but also the toppling
> of the present regime. Any person who is responsible
> for the use of these weapons would be held accountable
> in the future.[73]

Apart from implicit toleration of Iraq's unconventional arsenal as long as it was not used against Americans, Baker's coupling of biologicals with chemicals was itself novel. There was little talk of an Iraqi biological arms capability until four months before the war. Previously, U.S. officials maintained that Iraq was conducting biological warfare research but had not developed actual weapons. In September 1990, however, the U.S. assessment changed. CIA Director Webster stated for the first time that Iraq had a "sizable stockpile" of biological weapons. His claim was backed by intelligence reports provided to the House Armed Services Committee.[74]

Three months later, in January 1991, military officials were nervously trying to provide assurance that U.S. troops would be protected in case of biological or chemical attack. Some of the troops were receiving inoculations against anthrax and botulinum toxin. They were also provided with antidotes for nerve agents and were equipped with masks and outerwear intended to protect them from both biological and chemical weapons.

In the area of biological warfare, however, medical defenses could be ineffective if Iraq released unexpected organisms. Even if the Iraqis used a known agent such as the anthrax bacillus, protection was uncertain. Colonel Ronald Williams, commander of Fort Detrick's Medical Research Institute of Infectious Diseases, acknowledged that "if a vaccinated person were exposed to an overwhelmingly large dose [of anthrax bacteria], the protection might not be effective." Moreover, antibiotics might not work against certain anthrax strains developed by Iraqi scientists.[75]

The essential reality was neatly summarized by *New York Times* reporter Malcolm Browne: "The nightmare of biological warfare has long haunted medical experts working for America's armed forces, and years of research have underscored the near impossibility of protecting troops against all possible biological agents."[76]

The Greatest Fear

United States military officials were so worried about Iraq's chemical and biological arsenals that they believed the weapons could enable Iraq to triumph. During the first 40 days of the war, coalition bombing decimated Iraq's military infrastructure including, presumably, its chemical and biological production facilities. Nevertheless, General Norman Schwarzkopf, commander of the coalition forces, remained apprehensive. He especially feared chemicals, and implicitly biologicals, as ground operations began in late February:

> I could conjure up a dozen scenarios in which the Iraqis would make victory extremely costly, and I reminded my staff: "You can take the most beat-up army in the world, and if they choose to stand and fight, you're going to take casualties; if they choose to dump chemicals on you, they might even win." In the past Saddam had used nerve gas, mustard gas, and blood-poisoning agents in battle; and even though he hadn't fired chemicals on Al Khafji [a town in Saudi Arabia near the Kuwait border], I was still expecting them when we launched our offensive. My nightmare was that our units would reach the barriers in the very first hours of the attack, be unable to get through, and then be hit with a chemical barrage. We'd equipped our troops with protective gear and trained them to fight through a chemical attack, but there was always the danger that they'd end up milling around in confusion—or worse, that they'd panic.[77]

A survey of troops who served in the Gulf War confirmed their special concerns about chemical and biological weapons. About 2500 troops responded to questionnaires distributed in the aftermath of the war, between April and July 1991. Conducted by the U.S. Army Research Institute for the Behavioral and Social Sciences, the questionnaire inquired into "human performance issues in command and control" during the war. Views were sought about a range of issues, including leadership, combat capabilities, intelligence, and weapon effectiveness.

Among the 46 questionnaire items, only one referred to chemical (and inferentially to biological) weapons. For that item, respondents were asked, "How disruptive to your operations was each of the following?" Eight possible disruptions were listed, and respondents rated them on a scale of 1 to 7, where 1 was the least disruptive and 7 the most. The average rating for "chemical threat" was 3.6. This rating was the highest among all the listed choices. The seven other disruptions and their ratings were: staff sleep loss and fatigue (3.5), unit sleep loss and fatigue (3.3), command post displacement (2.8), enemy indirect fire (2.3), enemy direct fire (1.9), enemy maneuvering on the battlefield (1.7), enemy deception operations (1.5).

Thus, although chemical and biological weapons were evidently not used, the threat alone was deemed more disruptive than any other listed possibility. If troops had actually been attacked with these weapons, disruption could have been incalculable.[78] Frightened, suffocatingly hot, Iranian troops a few years earlier had ripped off their masks in panic. Saddam's chemicals ravaged their bodies and their morale. Would coalition soldiers have responded differently from the Iranians? As noted earlier, the possibility that they would not have, petrified General Schwarzkopf.[79]

The brevity of the ground war and the minimal resistance by Iraqi forces largely explain the moderate ratings for any of the listed disruptions. A prolonged conflict might well have altered the impression. In any case, the chemical threat exceeded all other disruptions including enemy fire, maneuvering, and deception operations.

The survey does not inquire into the reasons behind the ratings for the chemical threat. How much was attributable to fear is not clear, or to the burden of wearing protective gear, or to the slowed mobility in areas of suspected chemical or biological threats. Or all three. Nevertheless, the responses underscore how disruptive the anticipation of a chemical or biological attack can be.

Worries about Saddam Hussein's chemical and biological weapons were graphically recounted by a journalist who covered the war. While accompanying U.S. Marines, Molly Moore recalled her own anxieties as she listened to an instructor describe the effects of a biological attack:

> With anthrax, the symptoms won't start for two or three
> days. The first thing you notice are little black dots
> appearing on your skin. You have hours or maybe only
> minutes to live at that point. You'll feel like you have a cold

congestion. Then you can't breathe. When you're that bad, you're probably gonna die.[80]

By the time the instructor finished lecturing, Moore said she felt waves of nausea, headaches, and lung congestion.[81] Her description of another experience suggested how terrifying a chemical attack could be.

About 1:00 A.M. on January 20, 1991, several U.S. tanks were perched in the Saudi Arabian desert. Some crew members dozed, while others gazed at allied planes streaking north to Kuwait. Sergeant McKee was on watch.

The receiver in McKee's ears crackled to life. "Gas! Gas! Gas! This is not a drill."

McKee's eyes bulged. He felt like he had instant night vision. His screams pierced the heavy air. "Gas! Gas! Gas! This is not a drill." . . .

Oh my God, it must be bad, Mathews thought as he flew out of his sleeping bag, trying to pull on his chemical suit and scrambling to get inside the tank. The night was pitch black. In his frenzy, he nearly knocked Delaney off the tank as he too scurried for the open hatch. . . .

Delaney's mouth voiced words his ears didn't hear. "Relax! Relax! Take your time!" He leaped inside the tank behind Mathews and slammed the hatch shut, just as he realized that he was clad only in his mask and underwear. It was too late. He couldn't risk going back outside for his chemical suit.

McKee was already inside the tank, mask pulled over his face and radio clamped over his head. The chaos inside Red One was repeated up and down the tank lines. The voice from the command post continued to shout. "This is not an exercise. Gas! Gas! Gas!"

Page's leg was twitching with uncontrollable fear. Calm down! he ordered himself. Oh my God. My mask. I need my gloves. What's going on? . . . .

Mathews's heart felt like it was beating a million miles an hour as he squeezed into his tiny driver's compartment. Suddenly he felt the first symptoms. His eyes started watering. He knew his body would go into spasms any minute. The anxiety was almost unbearable. He thought about the atropine needles in the bottom of his canvas mask bag. Oh God, could he really jab one of those horse-sized needles into his thigh when the time came? His chest felt so tight he thought he was going to collapse. He gasped for air. I've got to have that last breath before I die, he thought. But he couldn't breathe. The end was near.

Page, who got claustrophobic just putting on a chemical suit, felt the burning on his hand first. His eyes started stinging. The gas was seeping under his mask. Shit. He was sucking the damn stuff into his lungs. "Is my mask busted?" he shouted to anyone who would listen. "Can you see? Is my mask broken? Is it broken?" . . . .

The minutes dragged. It was not going to be a quick and easy death. It was going to be agonizingly painful, each new symptom torturing each man until his body succumbed to violent seizures and he lay twitching on the tank floor.

The radios sputtered and crackled again. "False alarm. Unmask." . . .

Shaky hands tugged the ungainly masks from sweating faces. . . . [H]ad it been real, quite a few Marines would have died out of ineptness and fear. . . . One Marine hadn't been able to find his mask, so he just sat down on a cot and cried, awaiting what he thought would be certain death.[82]

Moore's account is reconstructed from interviews, notes, and logs of the troops at the scene. The false alarm created moments of horror. Even then, the presumed attack was at night and in winter, when the heat of the desert did not add to the burden of wearing protective gear.

After the war, when UN inspectors examined Iraq's chemical facilities, they wore masks and other special outerwear. A review of their performance found that, "In hot desert conditions, inspectors may be constrained to no more than 15 minutes in a toxic area because of the debilitating effects of protective equipment."[83] Thus, even in the absence of an enemy threat, wearing protective gear can quickly become unbearable. The plight of Iranian soldiers who took off their masks and ran from chemically drenched battlefields becomes more understandable. What the reaction of coalition forces might have been in the heat of day can only be imagined.

The anxieties among American troops prompted by the brief chemical alarm were shared by an entire nation for weeks. When air-raid alarms sounded in Israel, every man, woman, and child put on masks, fearing that Iraqi missiles might be armed with chemical or biological agents. The threat from Iraq was in part a consequence of the lowered moral barrier against poison weapons. Now they were terrifying a whole civilian population. How the Israelis coped is the subject of the next chapter.

CHAPTER 6

Israel in Gas Masks

"By God, we will make fire eat up half of Israel if it tries to do anything against Iraq," Saddam Hussein proclaimed in April 1990. Addressing the general command of his armed forces, he had previously underscored Iraq's chemical capabilities in his speech. Saddam's remark about eating up Israel was widely interpreted as a threat to attack with chemical weapons.[1] His actions against the Iranians and Kurds, as well as statements from other Iraqi officials, provided ample reason for Israeli concern.

This was not the first time Israel was alarmed about chemical warfare. Fears about an Egyptian capability added to tensions during the weeks preceding the six-day war in 1967. In consequence, Israel obtained 20,000 gas masks from the German Federal Republic on June 2, three days before the war began. A few weeks after Israel's swift victory, the masks were returned unused.[2]

The Egyptian scare prompted Israel to equip some bomb shelters with chemical and biological weapons filters.[3] But during the next two decades, Israelis became complacent about the idea of a gas or bacterial attack. Even the Iran-Iraq war failed to arouse the general public, although military officials were paying close attention. After witnessing the Iraqi use of chemicals with impunity, and fearing a Syrian capability, the army tried to enhance civil defense measures. In 1988, gas masks were distributed on a trial basis to several civilians. The people remained passive, however, and the masks were soon lost or stored in forgotten corners.[4]

The public's attitude changed abruptly in 1990. After Iraqi officials explicitly threatened Israel with chemicals, the nation began to intensify defense preparations. Hospitals and schools conducted chemical warfare drills, and plans were drawn to distribute gas masks and other equipment to every household.[5]

When Saddam invaded Kuwait in August 1990, Prime Minister Yitzhak Shamir said he expected war to spread to Israel. Civil defense officials advised citizens to purchase a two-week supply of food, as well as masking tape to seal their windows in case of a gas attack. Israeli television, radio, and newspapers began transmitting defense instructions. Although focused on the chemical threat, official statements later made clear that the proposed measures were intended to protect against biological weapons as well.

Underground bunkers would not be suitable, according to the government, because poison gas is heavier than air and settles on the ground. People were advised to seek safety as high above ground as possible.[6]

The government held millions of gas masks at the ready along with antichemical weapons kits containing antidote syringes against nerve gas, detection litmus paper, and decontamination powder. The public was jittery. The people who produced the protective gear, for example, said they were struggling to reduce their own fears. "Contributing to the defense effort helps us stay calm," said the factory's general manager.[7]

Toward the end of 1990, Israelis felt increasingly isolated. Every major airline except Israel's El Al had suspended flights to Tel Aviv–Jerusalem. Tourism virtually stopped. Overseas organizations canceled meetings they had long planned to hold in Israel. The only exception to the near freeze on entry was the continued immigration of Jews from the Soviet Union. (Even during the war, when Iraq was firing missiles into Israel, hundreds of Soviet immigrants arrived each day.)

Meanwhile, Iraq offered no indication that it would heed UN demands to withdraw from Kuwait by January 15, 1991. As the deadline approached, Israeli concerns about chemical or biological attack were mirrored in the mounting number of press features on the matter. In the weeks before the deadline, masks and kits were distributed to every citizen.[8] Plastic-covered cribs with filtered ventilators, called *mamats*, were provided for infants and small children. A veterinary supplies firm was overwhelmed with demands for specially ventilated tanks that could protect pet dogs and cats.[9]

The Prospects of Chemical and Biological Attack

From the first day of the Gulf War to the last, Israelis worried that Iraqi missiles might carry chemical or biological agents. Never before had an entire nation had to don gas masks. This extraordinary requirement framed the mood of the country, the assumptions of its leaders, and attitudes among individual citizens. The experience highlights the anxiety associated with the expectation of attack, even when none materializes.

The following narrative is drawn largely from interviews and from newspaper articles, advisory columns, and letters to the editor during the war. They highlight people's reactions by way of preparation, confusion, tragedy, even humor.

ADVICE

As the January 15 deadline neared and war seemed likely, Israeli media and civil defense officials offered a river of advice. People were told what to do in case of attack, what to expect, how to keep themselves and their children calm. A typical article, in the January 11 *Jerusalem Post*, posed worrisome problems and how they might be addressed. Presented in a question-answer format, the article was titled: "Q: What If There's War? A: You'll Be All Right." The first part of the article sought to address psychological distress:

> Q: I've been having trouble sleeping lately; I close my eyes and see Scuds. . . . [W]hat are the chances of survival for my family and me?

A: Even with minimal preparation, extremely good. . . .
[I]n the unlikely event of a chemical attack, extensive
military research has shown that your chances of survival
are markedly increased just by going indoors and upstairs
(the chemicals tend to sink). Go upstairs and seal an inner
room, and the survival rate skyrockets. Do the above and
properly don a gas mask . . . and the probability of survival
verges on 100 percent.

Q: My children aren't sleeping so well either. What's the
best way to calm their fears?

A: Experts suggest that the first step in helping children
with their fears is to make sure you are attending to your
own. Talk to other adults about what you're feeling. Nearly
everyone is in the same emotional boat. . . . As for your
kids, listen to them, answer their questions, ask about civil
defense preparations they've experienced at school, and
encourage smaller kids, especially, to bring out their
feelings by drawing pictures and telling you about them. . . .
Say to yourself, and mean it, "We really are going to be
fine."

The article then provided instructions for action:

Q: I want to prepare, but I'm confused. . . . Where do I
start?

A: A good place to start is to choose a room in your home
to seal in the event of a chemical warfare alert. . . .
[C]ollect materials needed for sealing both that room and
the outside openings of the apartment. These include rolls
of wide adhesive tape . . ., sheets of thick plastic (to give
another layer of protection in case of cracking), lots of
cloths or towels to seal the bottoms of doors, a bucket and a
small bottle of bleach. In the event of an alert, a little
bleach can be added to a bucket full of water, then the
towels can be soaked in the solution and placed against the
door bottoms.

(Although bleach was thought able to neutralize nerve gas,
it was later deemed unnecessary and dropped from advisories.)
The article suggested keeping several items available: waterproof

clothing and boots for each family member, drinking water in sealed bottles, flashlights, candles and matches, a transistor radio, a first-aid kit, canned and otherwise sealed foods, family games such as bingo and cards, toys, and books.[10]

A follow-up *Jerusalem Post* article, titled "What to Do If an Alert Is Sounded," appeared on January 14. It reiterated instructions in the earlier piece, with particular attention to the needs of children. It warned parents to expect youngsters "to show regressive behavior, to return to habits and actions (thumb-sucking, etc.) that you may not have seen for awhile." Children "may also show tendencies to be aggressive, or be exceptionally withdrawn and passive." The article advised patience, giving children attention, and helping them express their feelings.

"You're all doing the best you can with a most unnatural reality," the article tried to assure. It concluded with an emotion-tinged question: "Finally, my worst fear. What happens if I am separated from my children during an attack?" The answer was that children at school will be protected, and temptations to get them should be stifled—"you don't need them to be orphans."[11]

The strenuous efforts to provide assurances reflected the anxiety that gripped the country. Advice was offered about almost any contingency. Some recommendations seemed exotic, even frivolous, such as what to do if an attack occurred when driving a car:

> [D]rive as rapidly as you can, while wearing your mask, to the summit of the nearest hill—both mustard and nerve gas are heavier than air and sink to low-lying areas. At the top of the hill, you should wait in the closed-up auto (wearing your gas mask and covered by a sheet of plastic) until you feel it necessary to open a window. By that time, the gas cloud should have passed.[12]

The advice seems an invitation to chaos: anxious people driving as fast as possible to find hills in unfamiliar territory. All this while wearing masks that limit their vision. The risk of accidents would seem enormous.

Misperceptions

Advisories generally minimized the risks of chemical or biological warfare to properly protected civilians. But some prewar assurances proved dramatically wrong as soon as the scud attacks began. On January 14, for example, Israelis were told they would have plenty of warning time:

Q: How much advance warning are we likely to have before an actual attack begins?

A: As much as several hours or more. In the first stage of a low-level alert, announcements will be made on radio and TV and by roving loudspeaker vehicles.[13]

Four days later, the first scuds fell on Israel. Advance warning was less than two minutes. The flight time from Iraq turned out to be seven minutes, but the first five minutes was spent locating the scuds by U.S. satellites and confirming their trajectories. Experience quickly replaced wishful thinking. Under optimal conditions, according to a news article, Israelis could expect "a formal attack warning notice about 90 to 120 seconds before impact."[14]

Another important advisory proved wrong as well. Before the war, Brigadier General Yehuda Danon, chief of the Israel Defense Force Medical Branch, assured civilians that they would spend "only a fraction of an hour, perhaps only minutes" in a sealed room during an alert. He added that the recommendations for protection against a chemical attack would be equally effective against a biological attack.[15]

An advisory in the middle of war, and with the benefit of experience, negated the general's prewar premise about time in a shelter. It suggested that sealed rooms be large enough to provide occupants with oxygen "for at least three hours."[16] Some alerts did last only minutes. But during a few scud attacks, as recounted in the next section, people spent hours in their sealed rooms.

Another question-answer piece, on January 15, was intended to minimize anxiety. "How do I know that I'm not going to die in some chemical attack?" The answer was cloaked in transparent exaggeration:

Experts say that with the most minimal of preparations, you stand a much greater chance of being struck by lightning than being at all affected by a chemical attack. Make a small effort, prepare a sealed room, don your mask when instructed and your chances of survival are assessed at virtually 100 percent.[17]

One cannot fault these articles, many by *Jerusalem Post* reporter Bradley Burston, for trying to keep citizens calm. In truth, however, no one knew how effective the masks and sealed enclosures would be or how long people could function in their protective environment. Fortunately, the assumptions about "100 percent" protection never had to be tested.

A major issue that festered through the war related to the type of shelter best able to offer protection. Initially, the government urged that everyone seek safety in a sealed room, preferably at a high level. Some people demurred, however, and government statements themselves became confusing.

The issue turned on whether explosives posed a greater threat than chemicals or biologicals. "There are different kinds of attacks, right?" asked a man during a civil defense call-in show on the first day of the war. "In case of attack, how do we know whether to go down to the bunker or up to the safe room?"

The fumbling answer by the civil defense official on the radio line spoke to the incongruity of the problem. There was no good answer. But he tried:

> Today we see all possible attacks—of course we hope there
> won't be such a thing—but, God forbid, if such a thing
> happens, we are relating to it as a chemical warfare
> material attack, and thus the proper action is to close
> oneself in a sealed room.[18]

As the scuds began to fall, a few people ignored official advice and devised their own techniques. When a siren signaled, they entered their ground-level shelters. After hearing an explosion, presumably from a landed scud, they rushed up to their sealed rooms. In the second week of the war, a man was killed from falling debris as he tried to change shelters.[19] But debate about optimal locations continued through the war.

On February 11, the Civil Defense Force (sometimes referred to as the Home Front Command) allowed that an underground shelter was acceptable if sealed and reachable in two minutes. The next day, then former defense minister Yitzhak Rabin told a radio interviewer that he was not using a sealed room at all.

Rather, he went to an unsealed underground shelter in his Tel Aviv building, eight flights down from his apartment.[20]

Some of the proposed solutions, like those in a February 17 article, seemed intended more to mollify than to offer substantive help. By that time, 30 scuds had been fired at Israel, all with conventional warheads. People were becoming more concerned about explosive impacts than chemical or biological exposure. The article proposed that sandbags be stacked against the windows of ground-floor sealed rooms. For rooms on upper floors, "it might be possible to use smaller, tightly sewn bags to protect windows from inside the room."[21]

Uncertainty about Iraqi intentions was illustrated by conflicting newspaper articles throughout the war. On January 21, Amatzia Baram, an Israeli expert on Iraq, maintained that the Iraqis were unlikely to use chemical or biological weapons.[22]

Four days later, other Israeli experts emphasized Iraq's capability to launch chemical warheads. Their worries were compounded by the introduction of Patriot interceptor missiles. If a chemically tipped scud were hit by a Patriot, it might break into hundreds of shards and disperse the contents in midair. As the material floated to the ground, it could cover wide areas and decontamination "could be long and difficult."[23]

On January 28, Israelis learned that U.S. Secretary of Defense Richard Cheney believed there was a "distinct possibility" that Saddam Hussein "may eventually fire scuds with chemical warheads."[24] At the same time, an article indicated that "there have been a number of reports that Iraq is planning a biological warfare attack on both non-combatant Israel and the U.S.-led alliance based primarily in Saudi Arabia."[25]

The Scuds Arrive

At 2:00 A.M. on January 18, air-raid sirens woke the small country of Israel, which is about the size of New Jersey. The rolling sound signaled a chemical warfare alert that prompted four and a half million Israelis to don gas masks and hurry to safe rooms. With this alert, one day after coalition forces began air attacks against Iraq, the worst day of the war began for Israel.

The Worst Day

The first missile fell on Tel Aviv minutes after the siren sounded. Seven more followed—another in Tel Aviv, three in Haifa to the north, and three in open areas between the two cities. During the next day, three false alarms were followed by an authentic alert during which two more scuds landed in Tel Aviv.

Chemical and biological identification teams were rushed to the sites after impact. A driver of one testing unit saw a low cloud above the location to which he was heading. Convinced that the cloud was a chemical agent, he fainted en route.[26]

Another team actually found poison gas after a scud landed, although the public was not told about the finding until after the war. A postwar news account indicated that a "tense, prolonged examination" followed the discovery. The source eventually was determined to be industrial chemicals from a textile factory that had been hit.[27] But for an undisclosed period, Israeli military officials along with the general public believed the country was under chemical attack. (CNN and other networks initially reported that the scuds contained chemical agents.)

Despite an all-clear signal some 90 minutes after the first attack, several people remained in gas masks for hours. A few did so out of fear, others not understanding that the alert had passed. The second series of missiles followed the first by 30 hours. The intervening false alarms meant that Israelis were in gas masks and sealed rooms much of that period.

The first missiles caused property damage, but surprisingly few people were injured, and none directly killed as a result of impact. There were other tragedies, however. Three elderly women died of suffocation because they had not removed the seals from their gas-mask filters. A three-year-old girl was suffocated to death as her parents struggled with her to put on her gas mask.[28]

News accounts revealed the confusion and terror that gripped Israelis in the early attack. Panicked by the alert, one family wore masks for three hours and repeatedly injected themselves with atropine, the nerve-gas antidote. Atropine, which was included in the chemical warfare kits, can be toxic in excessive amounts.[29]

Another Israeli, Yitzakh Malkian, was sure that he and his family were in the midst of a chemical attack when a missile slammed into their Tel Aviv home. "There was concrete, explosive material, dust, everywhere," he told a reporter. "My neighbor was thrown on top of me, and when I got up, I saw there was no house left, that we had no sealed room left." A reporter recounts more of Malkian's story:

> Five of Mr. Malkian's brothers put on gas masks, but two others were left without protective gear because of the explosion. The men began arguing over who should wear the masks they had.
>
> "I told them, 'I've already breathed the poison, I'm going to die, so it makes sense that I go out and look for the masks that were lost in the explosion,'" Mr. Malkian recalled.
>
> He and another brother put rags soaked in water and baking soda over their faces and searched futilely in the dark for the masks.[30]

The days immediately after the first attacks were especially frightening. Saddam had demonstrated that his missiles could hit heavily populated areas, and Israelis still worried that a chemical or biological shower could be imminent. Despite newly installed Patriot batteries, the rate of success in intercepting incoming scuds was uncertain. When and where the next scuds would fall kept the nation on edge.

"I am an elderly person living alone. I am becoming very frightened. What can I do?" The question was posed in the continuing newspaper advisories intended to calm the nation. The questioner was told to call a hotline number for emotional support. Further: "Make arrangements to spend this period with others, if you can. Stay busy in your home doing small projects like fixing things, rearranging family albums and writing or tape recording memories for other family members."[31]

HOSPITAL PREPARATIONS

In the second week of the war, Dr. Dror Harats, a young physician and expert in chemical and biological defense at Jerusalem's

Hadassah hospital, discussed the implications of an attack.[32] A missile carrying conventional explosives might cause 100 injuries, he estimated. A chemically tipped scud could cause 1000 casualties and, despite treatment, 100 might die. These figures were speculative, he conceded. Moreover, gas would be affected by such weather conditions as temperature, wind, and rain, which could cause it to dissipate or to linger.

The consequences of an attack with biological weapons were even less certain. But Dr. Harats believed that the effects of a biological attack—anthrax in particular—could largely be countered with antibiotics. Israeli hospitals were stocked with antibiotics that medical personnel could only hope would prove effective.

The parking area behind the Hadassah hospital was filled with cars. Dr. Harats said that in case of biological or chemical attack, the lot would be emptied in minutes to make room for buses and trucks filled with victims. The victims would rapidly be divided between those who could walk and those who needed to be carried. Dozens of metal stretchers were piled next to the lot. Hospital aides in gas masks, boots, gloves, and protective clothing would strip the victims naked.

Cold weather or not, the naked patients were to be watered down under one of 70 outdoor showers that lined the path to the hospital entrance. Once inside, bleaching powders would be administered to exposed skin, and intravenous connectors would feed antidotes or antibiotics in hopes of counteracting the effects of the poisons. Speed was essential. Biological agents can cause irreversible damage in hours, nerve gas in minutes.

Once before, Jews had been attacked with poison gas, another Israeli observed, and the current symbols were horribly reminiscent of that experience. Upon arrival at Auschwitz in World War II, people were divided into two groups—the weaker ones and the more able. The weak were immediately stripped naked and herded into sealed chambers, where they died in minutes as lethal gas flowed from fake shower heads. The corpses were then burned in oversized ovens. Saddam's allegorical threat to turn Tel Aviv into a crematorium was not lost on the Israelis.

A renowned scientist, who during the scud attacks oversaw chemical defense in a region of Israel, stated at the time what every Israeli seemed to believe: Auschwitz would not be revisited. "Saddam Hussein will be able to use chemical or biological weapons against us only once." Implicit was an understanding that Israel would unleash a nuclear response.

LIFE GOES ON

A month into the war fewer scuds were falling, and none had carried chemical or biological agents. The country was more relaxed. Having been closed during the first weeks of the war, schools had reopened. Several days now passed between air-raid alerts, and many people had resumed routine activities. But everyone still carried a cardboard box containing a gas mask at all times. When the siren sounded, the whole nation put on masks and entered sealed rooms.

Despite the lessened anxiety, the threat of missiles, gas, and bacteria left signatures on everyday experiences. They reached to social and cultural activities as well as the cycles of life. An artist's exhibit in late February, for example, was titled *Paintings from a Sealed Room*. The displayed works, by Gad Ullman, were described by a reviewer as containing "signs, signals, and cyphers from the Gulf War." The reviewer took special note of the "ubiquitous gas mask" and other war artifacts that infused the artist's paintings.[33]

In a show of solidarity with Israel under siege, violinist Isaac Stern and conductor Zubin Mehta flew from the United States to perform with the Israel Philharmonic. In the middle of a concert on February 23 in Jerusalem, the air-raid alarm went off, and the orchestra left the stage. As instructed, the audience put on gas masks and remained seated. (By that time in the war, keeping an audience seated rather than sending it out to find sealed rooms seemed a prudential risk.)

Moments later, Stern returned to the deserted stage to play a Bach adagio. Newspapers carried his picture, wearing a gas mask while playing. The picture was taken during a rehearsal—at the performance he played without a mask. But everyone in the audience wore theirs until the all-clear signal. One can hardly conjure a more poignant image of determined civility in the face of barbarous threat: a theater filled with finely dressed people in gas masks, enraptured by the sounds of one of the world's premiere concert artists.[34]

Judy Siegel-Itzkovich described becoming a mother during the scud attacks. She was among 8000 Israeli women who gave birth at the time. By the time her son was five days old, he had twice been placed in a protective *mamat*. Who would have thought, she wrote, "that a collapsible aluminum-and-plastic monster that protects against nerve gas and anthrax germs would be as important a part of his layette as blankets, flannel underwear and diapers?"

In an article titled "From a Sealed Womb to a Sealed Room," she told of her worries before delivering her baby and after:

> What would I do if I felt contractions in the sealed room? Would I have to give birth while wearing a gas mask? How could I cope with the demands of motherhood on top of worries about the war? How would I drag the *mamat* with me wherever I took the baby?

She recounted a scene in the hospital days after she gave birth. A missile alert had just sounded:

> The maternity ward had been completely insulated with plastic sheets and sticky tape, with only a timid, infrequent airing through a few of the large picture windows during the day. . . . As we had done at home some 20 times before, all the mothers pulled out their gas masks—an incongruous addition to their starched pink hospital gowns.
>
> Well-rehearsed, the nurses picked up each infant, pushing the beds one by one into the corridor and sealing the babies into the protective tents on the floor. Many of us had become used to this emergency routine at home, but going through it after delivery was eerie and nerve-wracking. . . .
>
> Women in labor in the delivery room and those in the recovery room after Caesarian sections were not exempt from the requirement to don the cursed black masks. Forever seemed to pass until Nahman Shai finally released us from our misery with the good news that the two Iraqi missiles had fallen harmlessly in unpopulated areas. . . .

As if on cue, the mothers raced to the nursery as the staff were quickly extricating each well-wrapped baby from his *mamat* and returning him to his bed, matching the names on its wrist bracelet with that on the card taped to the bed.

Despairing that "this crazy ritual" might soon be repeated, the author concluded with a wish for Saddam Hussein: "May you be rewarded with a crushing defeat. May your swords be turned into plowshares, and our *mamatim* into greenhouses."[35]

Humor

"Who Needs Dumb Jokes at a Time Like This?" read the first part of a February 11 *Jerusalem Post* headline, followed by: "We Do." From the outset, Israelis used humor to relieve their anxieties. Light-hearted anecdotes and silly stories relating to the war appeared regularly in the country's newspapers. Irreverent and corny, many can only be appreciated in the context of the time. One, for example, was based on the ubiquitous coverage of the war by CNN television reporters:

Q. A Jew, a Moslem and a Christian are in a sealed room with one gas mask between them when the siren sounds. What do they do?

A. The Jew runs up to the roof to watch the Patriots in action, the Moslem runs up to the roof to shout "Allah Akbar!" and the Christian runs up to the roof to file a live report for CNN.

A second was prefaced by the observation that it sounded better in Hebrew than in English translation:

Q. Why haven't we heard from the prime minister lately?

A. Because they forgot to take him out of his *mamat*.

Another story had Prime Minister Shamir, whose relations with President Bush were often strained, saying, "The last time the Jews listened to a bush, they wound up wandering through the desert for 40 years."

Quips such as these served an important purpose. "Some people believe that if the situation gets really dire, we'll stop telling jokes," said Yair Garbuz, an Israeli satirist. "That's not true. Sometimes humor is the only way to express your feelings."[36]

Advertisements also used humor to tie products to the war, as in an automobile rental company's cartoon of a car's windshield encased in an oversized gas mask. T-shirts were sold bearing slogans such as "I survived January 15, 1991, in Tel Aviv" or "Hard Rocket Cafe, Baghdad." Another shirt depicted primates at various stages of evolution, the last being a human being wearing a gas mask. It was captioned: "What went wrong?"[37]

The *Jerusalem Post* printed light-hearted anecdotes sent in by readers. A submission by Liat Collins appeared in a February 14 column called "War in the Fast Lane."

> Good habits are, it seems, as hard to break as bad ones. Sitting in a sealed room and suffering from flu, one Jerusalemite got into the habit of raising her hand to her gas mask filter every time she coughed, until a neighbor sharing the room pointed out that the mask should keep the germs in just as well as out.

Ines Hulte told about her parents in Argentina who watched live TV broadcasts of the scud attacks. She called them at the end of each raid to say she was safe.

> During one attack, I called with my usual report as soon as our area was released from wearing gas masks. But instead of relief, I got this from my mother, halfway around the world: "The all-clear hasn't sounded yet; go back to your sealed room."[38]

Some mailed-in experiences revealed sad irony: "Our granddaughter was presented with a dollhouse for her fourth birthday. She promptly set about creating a sealed room for the doll family."[39] Others lent advice, as from a reader whose mask clouded up when she put it on.

> [A] young woman soldier advised me to take a potato, cut it in half and rub it on the inside of the mask's eyepiece. Naturally, I exploded in laughter at the folksy, low-tech

solution. The soldier peered at me and said in all seriousness: "Why does everyone laugh when I say that?" PS: The solution worked![40]

Days after the war began, some people decorated the brown cardboard cartons that contained their masks. Dangling from shoulder straps, the boxes were ever-present, because Israelis carried them wherever they went. By the beginning of February, "designer carrying kits" were being sold in stores. Most Israelis continued to use the standard carton, but a few replaced them with illustrated polyethylene bags, multicolored nylon covers, or home-designed cases. In describing the fad, a reporter indicated that the idea was not entirely novel. She mentioned the recollection of her associate

> who experienced the blitz of Britain [and] said that it was customary during World War II for fashionable Londoners to have "four or five different gas-mask carrying kits to match various outfits."[41]

The Scud War Ends

Israel experienced the Iraqi gas and biological threat for six weeks. How its citizens would have behaved had the siege been longer can only be guessed. At the behest of the United States and other coalition countries, Israel remained out of the war. It agreed to suffer the scud attacks because of fears that Arabs in the coalition would not tolerate an Israeli military response.

The U.S. military command insisted that its air attacks were destroying Iraq's scud launchers and that Israel could add nothing to the effort. The goal of eliminating the launchers, according to General Norman Schwarzkopf, head of the coalition forces, was so dedicated that it distracted from other military efforts.[42]

The Israeli assessment differed sharply. Toward the end of the conflict, Israeli restraint was begrudging. The Israelis believed that allied planes were arriving at launch sites after Iraq had removed the launchers. They reportedly wanted to place their troops on the ground to locate and destroy the launchers.

Washington would not accede to Israeli requests for a safe air corridor, however. Moshe Arens, Israel's defense minister, complained bitterly about the U.S. refusal.[43]

After the war, Schwarzkopf conceded more uncertainty than he had formerly acknowledged about the elimination of Iraq's mobile missile launchers: "The launchers turned out to be even more elusive than we'd expected. We picked off a few, but just as often bombers would streak to a site where a missile had been launched only to find empty desert."[44] Arens's assessment was more severe: During the six-week war, the allies "did not manage to destroy a single Iraqi mobile missile launcher."[45]

If scuds had continued to fall, Israel probably would have entered the war despite American misgivings. It was on the verge of doing so when President Bush ordered a cease-fire on February 28.[46] If chemical or biological weapons had been used, Israel almost certainly would have retaliated immediately. Its leaders had repeatedly promised to respond to such an attack with overwhelming force.

The last missile fell on Israel February 25, and the war was over three days later.

A national survey indicated that the primary source of fear during the war was uncertainty and the second was gas.[47] The psychological damage, however, was not quantifiable. Just before the war ended, Gila Svirsky lamented about its effects on young people as she recapitulated her own family experience:

> Denna, my 15-year-old daughter, has not slept well since August. The seven months of preparing for the war and then war itself have been terrifying for her. It's no wonder that her nerves are frayed and rent anew with every siren. She has helped seal our house, has seen me purchase long term provisions and put them into glass jars, has helped install emergency lighting, has stored a spare set of her clothes in plastic bags, has learned how to use a gas mask and a hypodermic antidote against nerve gas, and, finally has gone racing into the sealed room as 39 Scud missiles fell. Not surprising that she dreams of war and wakes up crying.[48]

A study showed that during the war, Israelis commonly suffered from insomnia, many because they feared going to sleep.

They worried about not hearing the air raid alarm, or awakening in a state of confusion that could delay putting on their masks and entering a sealed room. Whether awake or asleep, the population evidently was consumed by the issue. A survey toward the end of the war revealed that the most frequent subject of dreams had become the gas mask.[49]

How permanent the effects of these experiences on people's psyches is not known. Psychological injuries are not included in casualty counts.

During the war, 39 scuds landed in Israel. More than 4000 buildings were damaged and some 1600 families displaced. One thousand people were injured. Two persons were killed from direct scud hits, and 18 from indirect causes such as heart attacks and suffocation from improper use of gas masks.[50]

A 1995 study suggested that indirect casualties were much greater than reported previously. On the first day of scud attacks, death rates rose throughout the country by 58 percent. On January 18, 1991, 93 deaths would have been "expected," but 147 were recorded. The increases were most frequently caused by heart attacks, presumably from stress, and respiratory complications, probably related to gas-mask problems or poorly ventilated sealed rooms.[51]

Citizens also learned after the war that there was far less certainty about how they would have fared in a chemical or biological attack than they had been led to believe. The masks distributed to the public were a single size. State Comptroller Miriam Ben-Porat cited army documents suggesting that one size could not be adapted to all facial contours. Leakage would have occurred in as many as one-third of the cases, according to the documents. She blamed the Defense Ministry for insufficient budgeting and advance preparation. The army disputed her conclusions about the inadequacies of the masks.[52]

Was the comptroller correct? Sometime after her report, the Home Front Command began placing notices in newspapers. It emphasized that "only a properly fitted mask ensures your safety!" Locations throughout the country were then listed where new gas masks and fittings were available. The notices urged "replacement of old gas masks for all who have Gulf war gas masks which they have not yet changed."[53]

In the end, the Israeli experience offers several lessons about mounting a broad civilian defense program against chemical and biological weapons. First, until the Iraqi threat seemed imminent, the public showed little concern about chemical and biological warfare matters. Army efforts in the 1980s to enhance civilian chemical defense were met with indifference. Second, late in 1990, when an Iraqi chemical and biological attack seemed increasingly possible, the government and media began saturating the public with warnings and advisories. Only then was the public roused.

Third, given weeks of advance notice, the civilian population was able to prepare in some measure for chemical or biological attacks. But, fourth, even in this highly educated and motivated society, confusion about proper defensive actions continued through the war. Finally, the civilian defense measures probably benefited the general morale. But how effective they would have been against an actual chemical or biological attack remains speculative.

CHAPTER 7

Unanswered Questions

Who could have guessed that five years after Saddam Hussein's overwhelming defeat in the Gulf War, he would still be in power. George Bush had long since lost his presidency despite brilliant orchestration of the war. But in 1995, Saddam's lock on Iraq was reaffirmed in a presidential election giving him 99.7 percent of the vote. (The lopsided margin was doubtless enhanced by the absence of an opposing candidate.)

Although Saddam retained his office, however, the principal goals of the Gulf War had been achieved. Kuwait was independent, and Iraq was no longer a serious military threat. Moreover, as the coalition promised during the war, sanctions would continue until Iraq eliminated its nuclear, chemical, and biological warfare programs.[1]

But three large questions concerning biological and chemical warfare continue to linger after Iraq's defeat. The first is why Iraq apparently did not use these weapons in the conflict. The second relates to thousands of veterans who claim that they became ill as a result of the war, possibly from exposure to Iraqi chemical or biological agents.

The third unresolved issue concerns Iraq's obligation to eliminate its unconventional weapons. Verification of Iraqi compliance has not been easy, and UN monitoring efforts offer lessons for other arms control agreements as well.

Iraq's Failure to Use Chemical or Biological Weapons

Many veterans of the Gulf War, and a few members of Congress, are convinced that Iraq used chemical or biological weapons during the conflict. The evidence is unclear. Chemical agents more than biologicals are amenable to field detection, and sensing devices were available in the battle areas. In fact, detection teams from two coalition countries found traces of nerve and mustard agents during the war.

The report of a Czech unit that discovered chemical agents was later deemed valid by U.S. Defense Department experts. At the same time, Secretary of Defense Les Aspin said the United States continued to believe that Iraq had not used chemical weapons. He suggested that the Czech readings may have been caused by "some kind of industrial chemical."[2]

A French military officer informed the U.S. Senate Armed Services Committee that French troops had detected low levels of chemical agents during the war. Responding to the French findings, Senator Richard Shelby said: "It further confirms to me that our troops and our allies were exposed to chemical warfare agents."[3]

The next section, on Gulf War syndrome, deals with the possible presence of such agents. But even if trace amounts were identified, no large-scale use of chemical or biological weapons occurred. This in itself surprised coalition military planners.

The troops arrayed against Iraqi forces had good reason to expect such attacks. Iraq used chemicals extensively against Iran and was believed to have subsequently expanded its chemical and biological arsenals. Moreover, Iraqi leaders threatened to use them. After the Gulf War began, Saddam Hussein told a CNN reporter that Iraq could attach nuclear, chemical, and biological arms to its missile warheads. "I pray to God I will not be forced to use these weapons," he said, "but I will not hesitate to do so should the need arise."[4] Why didn't he?

THE RANGE OF CONJECTURE

The earliest authoritative conjecture came from General H. Norman Schwarzkopf. At a February 27 press briefing, one day before the war ended, a reporter asked the Desert Storm commander: "Is there a military or political explanation as to why the Iraqis did not use chemical weapons?" Schwarzkopf had no doubt that they had not been used and seemed horrified about what would have happened if they had been:

> We've got a lot of questions about why the Iraqis didn't use chemical weapons, and I don't know the answer. I just thank God they didn't.

The reporter wondered if "they didn't use them because they didn't have time to react?" The general responded at length, concluding again with thanks to God.

> You want me to speculate, I'll be delighted to speculate. Nobody can ever pin you down when you speculate.
>
> Number one, we destroyed their artillery. We went after their artillery big time. They had major desertions in their artillery, and . . . that's how they would have delivered their chemical weapons, either that or by air. And we all know what happened to their air. So we went after their artillery big time, and I think we were probably highly, highly effective in going after their artillery.
>
> There's [*sic*] other people who are speculating that the reason they didn't use chemical weapons is because they were afraid if they used chemical weapons there would be nuclear retaliation. There are other people that speculate that they didn't use their chemical weapons because their chemical weapons degraded, and because of the damage that we did to their chemical production facilities, they were unable to upgrade the chemicals within their weapons as a result of that degradation. That was one of the reasons, among others, that we went after their chemical production facilities early on in the strategic campaign.
>
> I'll never know the answer to that question, but as I say, thank God they didn't.[5]

A year later, speculation about Iraq's chemical/biological restraint was still percolating. A staff member at the army chemical school at Fort McClellan, Alabama, did not believe the question would soon be settled.[6] In her book about the war, journalist Elaine Sciolino pondered what she called the "great mystery," and largely repeated Schwarzkopf's conjectures.[7]

We may never learn the complete answer, but some explanations seem less plausible than others. Two commonly posed reasons—that Iraqi chemical agents had become degraded or that coalition bombings had destroyed the arsenals—are improbable.

In April 1991, Iraq gave the United Nations a list of its weapons of mass destruction as required by Security Council Resolution 687. The inventory included some 10,000 nerve-gas warheads, 1000 tons of nerve and mustard gas, 1500 chemical weapons bombs and shells, 52 scud missiles, and 30 chemical and 23 conventional warheads.[8]

Formidable as the inventory was, U.S. officials believed it was incomplete. Indeed, in July, Iraq acknowledged it had more than 46,000 chemical munitions, and even then the UN inspectors believed there were more.[9] All this *after* the war, when much of the arsenal had supposedly been destroyed. Inspectors later determined as well that Iraq's ability to use the nerve agent sarin as a weapon was intact and that its mustard agents remained "good and usable."[10]

Similarly, the notion that the coalition forces moved too quickly on the ground for the Iraqis to react seems unlikely. An allied commander said that the Iraqi army was "caught off guard" and did not have time to organize a chemical attack.[11] But the ground assault was no surprise. Saddam may not have known the exact date, but he knew an attack was imminent, as did everyone else in the world. His troops could have had chemical or biological weapons at the ready if so ordered. Virtually none were found near the front lines.

Moreover, Iraq had ample opportunity to use chemicals or biologicals against Israel and Saudi Arabia in the weeks preceding the coalition attack. The postwar inventory included scud missiles with chemical warheads. Even more frightening, Iraq later admitted that its missiles had also been armed with biological agents.[12] But the scuds fired by the Iraqis bore only conventional warheads.

To suggest that Saddam refrained from using biological or chemical weapons because he believed coalition forces were defensively prepared also seems questionable. Saddam's use of gas against the Iranians was instructive. He and his military advisers must have understood the fears of coalition commanders. This was especially true in the psychological realm. Troops in a chemically or biologically poisoned environment might well have become terrorized and ineffective despite gas masks and outerwear. That is precisely what happened with the Iranians.

FEAR OF RETALIATION

The most likely reason for Iraqi restraint appears to be fear of retaliation. Had the Iraqis used these weapons, they almost certainly would have received a devastating response. This they were promised on several occasions.

Early in January, President Bush told Saddam Hussein that "the United States will not tolerate the use of chemical or biological weapons, support of any kind of terrorist actions, or the destruction of Kuwait's oil fields and installations."[13] In the matter of the oil fields, Iraq ignored the president's threat. Hundreds were set ablaze, and they burned for months. Thus, fear of the president's admonition was not itself sufficient to preclude Hussein's defiance. But using biological or chemical weapons was categorically different from oil fires. (Iraqi support of terrorist efforts against U.S. targets has been alleged by an unidentified analyst.[14])

Subsequent warnings by American officials about the use of chemicals and biologicals stood on their own. They were not linked to oil or other matters. Thus Saddam had good reason to expect a more drastic response to the use of chemical or biological weapons than to other actions.

In his meeting with Iraq's foreign minister, Tariq Aziz, on January 9, 1991, Secretary of State James Baker pledged that the use of such weapons against American forces would mean the end of the Iraqi regime. Furthermore, persons responsible for their use "would be held accountable in the future."[15]

When Iraq was firing scuds into Israel, U.S. Secretary of Defense Richard Cheney warned that if the Israelis were attacked with chemical weapons, they might retaliate with weapons of mass destruction.[16] As noted in Chapter 6, though Israelis would not specify the means, they made clear that their response would be overwhelming.

All this must have impressed Saddam. Iraqi reluctance to attack Israel or coalition forces with chemical or biological weapons seems to accord with fear of retaliation. This was borne out in conversations with officials from several countries.[17]

According to Donald Mahley of the U.S. Arms Control and Disarmament Agency, Iraq understood that using chemicals "might very well have altered our war aims to their disadvantage." This was Mahley's euphemism for what Bush and Baker had bluntly stated at the time of the war: Iraq would have suffered punishing retaliation.[18]

Graham Pearson, former director general of Britain's Chemical and Biological Defense Establishment, also was blunt: "Saddam Hussein was warned by Baker that if he used unconventional weapons, the United States would not rule out a possible response with unconventional weapons."[19] Baker did not directly say the United States would use unconventional weapons. In fact, President Bush had ruled out the possibility in December. But Baker purposely left a different impression with Tariq Aziz. In his memoir he says he implied that an attack with chemicals or biologicals "could invite tactical nuclear retaliation."[20]

If Saddam could not prevail against the coalition forces, he at least wanted to survive. According to Freedman and Karsh, Saddam was convinced he could do so as long as he did not use chemicals or biologicals.[21] Wrong about so many other assumptions, Saddam nevertheless proved right on this one.

Indeed, Iraqi officials eventually admitted that they had feared retaliation. In August 1995, they told UN inspectors that Iraq did not use these weapons because the United States warned that such action would provoke a devastating response.[22]

A final word about Iraq's failure to use biological or chemical weapons: I have recounted the views of analysts and diplomats from countries with particular interest in the matter—the United States, Great Britain, Iraq, and Israel. They offered a range of opinions about why Iraq did not use these weapons during the Gulf War: lack of time to prepare, degraded agents, a disrupted communications system, destroyed stockpiles and delivery systems, perceived ineffectiveness, and, of course, fear of retaliation. But among the array of conjectures, the word morality was uttered by no one.

By the time of the Gulf War, Iraq's past behavior, and the world's silence about that behavior, had negated "morality" as an explicit reason for restraint. Nevertheless, the American threat of overwhelming retaliation implied that chemical and biological weapons were still seen differently from conventional weapons. No one suggested that Saddam's restraint was a function of morality or ethics, because on the surface it was not.

The coalition's retaliatory threats had become part of a drama whose rationale went unpronounced. The notion of conscience as a barrier had been weakened in the Iran-Iraq war. A worldly axiom was laid bare: lowering a moral threshold is easier than maintaining one.

But Iraq's fear about using its biological or chemical weapons in the Gulf War is also instructive. Rules, norms, and threat of force can prevent heinous actions. They are sometimes necessary to prevent immoral behavior.

Gulf War Syndrome

In August 1992, as part of the Persian Gulf Registry Program ordered by Congress, the Department of Veterans' Affairs established three health referral centers. Located in Washington, D.C., Houston, and Los Angeles, the centers were to evaluate unusual symptoms being reported by Gulf War veterans. They were complaining of rashes, fatigue, diarrhea, chronic cough, joint pain, memory loss, and more. The symptoms, which appeared singly or in combination, at first were considered psychosomatic or related to smoke from burning oil wells.[23]

Later reports suggested the veterans' problems may have been related to leishmaniasis. Produced by a parasite that is transmitted by sandfly bites, leishmaniasis is endemic in the Gulf area. The disease causes fatigue, diarrhea, and heavy sweating.

Symptoms may not appear for months after the parasites are introduced, and even then the parasites are difficult to identify.[24]

Army spokesmen said evidence of the parasite had been found in a few servicemen and -women. But hundreds with similar symptoms complained of inadequate attention by the Veterans Administration (VA).[25]

Congressional hearings underscored the frustrations of men and women who had served in the Gulf. In 1993, several recounted their experiences to the Senate Committee on Veterans' Affairs. Colonel Herbert J. Smith typified the problem. A veterinarian, he was in excellent health before going to the Gulf. There he developed a cough, swollen glands, and joint aches. Upon return to the United States, he began to suffer memory loss and "couldn't even walk in a straight line without help."

No one in the military provided a diagnosis, although an army neurologist told Smith that his problems were "related to old age." (Smith was 50 in 1991.) Having witnessed numerous dead animals covered with dead flies, Smith believed his illness was caused by exposure to some sort of chemicals. He criticized VA physicians for refusing to make a diagnosis. Their reluctance allowed the government to avoid responsibility for the disabilities that he and other veterans were suffering:

> [T]hey're admitting these signs and symptoms are very bizarre, that there is no known disease that causes these symptoms to occur, yet they refuse to admit the possibility of a multiple chemical exposure that might cause this.[26]

Smith and other veterans at the hearing gained the sympathy of several senators. This was clear from the senators' impatience with witnesses like Frances Murphy. The coordinator for the VA referral center in Washington, Murphy testified that no connection had been established between the illnesses and service in the war. Committee chairman John D. Rockefeller IV responded in disbelief:

> Dr. Murphy, you've seen, assuming you've been in the room, five soldiers this morning. Most talked about their physical condition before they went in, and all talked, I think highly credibly, about the condition they now find

themselves in medically, which, for the most part, is grave, highly perplexing, and enough to cause them not to be able to get health insurance, not to be able to have jobs, to lose self-esteem, and who knows what else.

So my question to you is, what do you make of that? I mean, were these just unusual people?

Rockefeller was hardly satisfied when Murphy answered that the people with undiagnosed conditions just needed further study.[27]

The list of possible causes now included infectious disorders, exposure to smoke, fuel, pesticides, or multiple chemical sensitivity. Moreover, some veterans reported that their wives were suffering from similar symptoms, and babies born after the war had unusual numbers of health problems.[28]

Initial conjectures about the cause of Gulf War syndrome did not include chemical or biological warfare agents. But the possibility gained currency in 1993. Newspapers began carrying accounts of servicemen who believed the agents were the root of their problems.[29] Some recalled being in areas where chemical alarms sounded, followed by a burning mist. Although told by commanders that the alarms were false, the veterans claimed they signaled the beginning of their mysterious illnesses.[30]

AGENT ORANGE REMEMBERED

By December 1993, thousands of veterans reportedly were suffering from what was being called Gulf War syndrome. The earlier Pentagon emphasis on psychological causes had changed. Although conceding the possibility of physical sources, however, officials rejected any relation to chemical or biological warfare agents. But they still had no idea what the cause was.

"Here we go again. This is agent orange revisited," said a veterans' spokesman in 1993.[31] The reference to Agent Orange must have prompted shudders in the Pentagon. The chemical was used during the Vietnam War to defoliate forests and croplands. Intended to deny hiding areas to North Vietnamese forces, Agent Orange was later cited by American soldiers as a cause of fatigue, nervous disorders, cancer, and other illnesses.[32]

The Vietnam troops translated anger into relentless criticism. The government was protected by law from court action by former service personnel. But the veterans entered a class-action suit against seven chemical companies that produced Agent Orange. In 1984, the companies settled out of court for $180 million. But the issue did not end. Two firms sued the government for reimbursement of their shares of the settlement, and in 1995 the U.S. Supreme Court agreed to hear the case.[33] Twenty years after the Vietnam War ended, the issue was still festering.

The Agent Orange experience was not lost on the Departments of Defense and Veterans' Affairs, and probably every other government agency. They recognized the danger of dismissing the concerns of the Gulf War veterans. By 1994, as Gulf veterans continued to complain, government agencies became more responsive. In May, for example, Secretary of Defense William Perry and General John Shalikashvili, chairman of the Joint Chiefs of Staff, called on veterans with problems to "please come in for a medical examination." They urged them to register with the defense department's surveillance system or the Veterans Administration Gulf War Health Registry.[34]

In November 1994, an unusual law was enacted. Gulf veterans would receive compensation based on their symptoms, irrespective of the cause. They had only to demonstrate that disability from a chronic illness began during Gulf War service or within two years after.[35]

Despite such efforts, many veterans remained dissatisfied. Following the Agent Orange example, they filed a $1 billion suit in June 1994 against 11 companies. They alleged that materials sold to Iraq were used to make chemical and biological weapons that caused them injury.[36] But confirming a connection between Gulf War syndrome and these agents has proved elusive.

A possible biological/chemical weapons component of the illnesses has been conjectured in two areas. The first is that such weapons caused the illnesses; the second, that pretreatment drugs or vaccines were responsible. The most vigorous advocate of the biological/chemical agent connection was Senator Donald Riegle, Democratic chairman of the Committee on Banking, Housing and Urban Affairs. His authority derived from his committee's jurisdiction over export administration. This included earlier sales to Iraq of chemical and biological materials that could be used for military purposes.

EXPOSURE TO CHEMICAL OR BIOLOGICAL AGENTS

Riegle became interested in the possible linkage of these materials to the veterans' illnesses in July 1993. The Czech minister of defense had announced that a Czechoslovak unit detected the nerve agent sarin in areas of Saudi Arabia early in the Gulf War.[37] Subsequently, Riegle and his staff wrote several reports suggesting that American exports contributed to Gulf War syndrome.

Riegle and Alfonse D'Amato, the ranking Republican on the committee, stated categorically that "U.S. forces were exposed to some level of chemical and possibly biological warfare agents during their service in the Gulf War." Their conclusion was based on the nature of the exports, the reported identification of chemical agents during the war, and statements by U.S. veterans.[38]

The committee staff, according to a May 1994 report, interviewed hundreds of veterans (the precise number is not stated). Several said that, at different times and locations, chemical alarms went off after explosions. People in these areas became ill soon after. The ailing veterans complained of memory loss, muscle and joint pain, intestinal and heart problems, fatigue, rashes, sores, and running noses.[39]

The May report listed several apparent sources of these symptoms. The first was exposure from "direct attack" with Iraqi chemical and biological agents. Other possibilities included fallout from bombed-out plants or storage facilities, anti-nerve-agent drugs, and contact with Iraqi prisoners of war.[40]

> Despite the Department of Defense's position that no evidence exists . . . this investigation is establishing that there is substantial evidence supporting claims that U.S. servicemen and women were exposed to low level chemical warfare agents and possibly biological agents and toxins from a variety of sources.[41]

Riegle's committee staff issued another report in October 1994 to buttress these views. It contained additional statements by veterans and by experts who analyzed their claims. Former sergeant David Allen Fisher, for example, brushed against crates marked with skulls and crossbones while searching a bunker in southern Iraq. Within hours, his arm reddened and developed painful blisters.

The incident was confirmed by a chemical defense officer and others at the scene. But a Pentagon spokesman later denied that crates were even present. "In this case, as in the other cases like it," Riegle's October report said, "it seems impossible to obtain an explanation from the Department of Defense that is consistent with the events as reported by the soldiers present."[42]

Afterward, Riegle accused the Pentagon of "deliberate misrepresentation" and suggested its actions were criminal: "To my mind there is no more serious crime than an official military coverup of facts that could prevent more effective diagnosis and treatment of sick U.S. veterans."[43]

No one has tried harder to make the case that chemical or biological agents were responsible for Gulf War syndrome than Riegle and his staff. But their argument rests largely on anecdotal evidence. Not that individual experiences should be ignored or that Riegle's accusations are demonstrably wrong. Nevertheless, the linkage of complaints with exposure to warfare agents remains speculative.

Riegle did not seek reelection when his term ended in 1994. After he left the Senate, the position he advocated seemed in limbo. Other public officials believed that biological or chemical warfare agents may have caused Gulf War syndrome. But none pursued the matter with Riegle's dedication.

PYRIDOSTIGMINE BROMIDE

Another possible explanation is the use of a pretreatment drug called pyridostigmine bromide. Intended to neutralize the effects of nerve agents, the drug was taken by troops in anticipation of an Iraqi chemical attack.

A nerve agent works by interfering with the transmission of nerve impulses. An understanding of the process explains the rationale for pyridostigmine bromide.

Actions of the body are prompted by messages sent through nerves. The process includes the release by nerve endings of chemicals called neurotransmitters. A neurotransmitter transfers an impulse from one neuron (nerve cell) across the synaptic space to another neuron or to a muscle cell. When released and then received by an adjacent structure, the neurotransmitter induces a physiologic response affecting actions such as breathing, heart rate, and muscle contraction.

The response quickly ends when an enzyme (located near the receptors of the postsynaptic cell) causes a breakdown of the neurotransmitter. Acetylcholine is a neurotransmitter, and the enzyme that causes it to break down is acetylcholinesterase. A nerve warfare agent binds to acetylcholinesterase and inhibits it from acting. When this happens, the acetylcholine continues to accumulate.

The nerve impulse is unable to shut off. The ceaseless neurological "command" may disable the lungs and heart or keep a muscle contracted. The resulting shutdown of the body's neurological functions can cause death in minutes.

Enter pyridostigmine. If administered before exposure to certain nerve agents—for example, soman—pyridostigmine bromide can temporarily bind to acetylcholinesterase and help block the agent from binding to acetylcholinesterase. Thus, pyridostigmine mimics the effect of the nerve agent, though without the potency or comparable danger to the body. Indeed, pyridostigmine pretreatment does not offer protection by itself. It enhances the effects of other counteracting drugs, such as atropine, taken after nerve-agent exposure.

Moreover, pyridostigmine is not effective against all nerve agents. Ironically, according to Pentagon studies, pyridostigmine pretreatment may make people *more* vulnerable to sarin and VX, which were in the Iraqi chemical arsenal.[44]

The suggestion that pyridostigmine could be a cause of Gulf War syndrome was highlighted in May 1994 during a Senate committee hearing. Under chairman John Rockefeller, the Committee on Veterans' Affairs heard testimony from veterans, scientists, policy analysts, and government officials.

Former lieutenant colonel Neil Tetzlaff recounted his experience with the drug. Before arrival in Saudi Arabia in August

1990, he received 21 pills, one to be taken every eight hours. Although nauseated and vomiting from the day he started the pills, Tetzlaff said a doctor in Saudi Arabia first thought his illness was related to something he ate.

"On the sixth day," he testified, "I was incoherent, extremely tired, and at times irrational. On the morning of the seventh day I vomited about a quart of blood." Tetzlaff was hospitalized and since then has suffered from fatigue, palsy, and memory loss. He is convinced his problems were caused by a chronic overdose of pyridostigmine.[45]

Tetzlaff said that nothing was noted in his medical records when he was taking pyridostigmine. Nor were other soldiers suffering from overdose symptoms told to stop taking the drug.

> I find it interesting that many Gulf War veterans are reporting mysterious illnesses of pain, chronic fatigue, and other nerve disorders, and neither the VA nor the DOD has asked any of them if they took pyridostigmine. Because no records were kept, no one knows who did or didn't take pyridostigmine.[46]

The Defense Department assumed the drug was safe on the basis of earlier studies of a few healthy men. But Rockefeller observed that 400,000 troops had taken the drug while stationed in the Gulf. An unknown number could have been sensitive to pyridostigmine, especially if receiving other medications at the same time. Moreover, testimony at the hearing indicated that, when combined with pyridostigmine, an insecticide could become ten times as lethal against cockroaches. Might such deadly synergism apply to people?[47]

Edward Martin, a Defense Department spokesman, conceded that pyridostigmine "continues to be an investigational drug." He noted, however, that it had long been used in greater dosage to treat myasthenia gravis, a neuromuscular disorder. He believed the decision in 1990 to administer the drug to the troops was "the best scientific/medical judgment that could be made."[48] What was the ethical alternative? he asked rhetorically, and then answered: "Given that there was a very distinct possibility of substantial nerve gas exposure by troops, the most ethical decision was to provide the protections that we did."[49]

If concern about protection was understandable, the Pentagon's failure to keep records was inexcusable. "Some soldiers took the pyridostigmine bromide for 2 days and some for 2 months," Rockefeller observed. "There are no studies of the safety of the drug for healthy people for more than a week or two." The perplexed senator asked: "If we find out that pyridostigmine or some other of these investigational drugs turns out to be the cause of some Persian Gulf diseases, how on earth are you going to know who received them?" To which Martin meekly responded: "For pyridostigmine, there would be no record."[50]

In December 1994, the majority staff of the Senate Veterans' Affairs Committee issued a 53-page report on military research and veterans' health. Included was extensive discussion of the possibility that pyridostigmine bromide was a cause of Gulf War syndrome. Because its effects presumably are reversible, the drug does not pose the extreme danger of a warfare agent such as sarin.[51]

But the report refers to a Department of Defense study that raises doubts about the agent's safety. Pyridostigmine was given to four healthy men in August 1990. From the committee report:

> It is important to note that this study, conducted just prior to the Gulf War, included extensive safety precautions, including giving medical exams to the men before giving the pyridostigmine. *The researchers indicated that pyridostigmine should **not** be given to individuals who had bronchial asthma, peptic ulcer, liver, kidney, heart disease, or hypersensitivity to pyridostigmine or related drugs* [original emphasis]. They informed study volunteers that possible adverse effects include nausea, vomiting, slow heart rate, sweating, diarrhea, abdominal cramps, increased salivation, increased bronchial secretions, and pupil constriction. They also warned of other side effects, including "weakness, muscle cramps, and muscle twitches."[52]

The adverse effects resemble the symptoms reported by ailing veterans. These effects, and others associated with overdosage of pyridostigmine bromide, are also noted in a standard medical reference book.[53]

The Senate committee report then criticized the ethics of using the drug on the troops.

> *In contrast to the extensive precautions taken before giving pyridostigmine every 8 hours for 3 days to **four** volunteers, a few months later approximately 400,000 U.S. soldiers were ordered to take the same dosage of the drug for days, weeks, or months, none of whom had been screened for any of the diseases mentioned* in the informed consent form given to the four men, none of whom were warned about the risks associated with the drug, and none of whom were given a choice of whether or not to take it.[54] [Emphasis in original.]

Apart from this drug, inoculations against biological agents also have been cited as possible causes of Gulf War syndrome. But the anthrax and botulism vaccines seem less suspicious than pyridostigmine. In part this is because fewer people were exposed—about 150,000 received vaccinations against anthrax and 8000 against botulism. Beside the larger numbers that took pyridostigmine, the drug's side effects make it a more likely candidate.[55]

In a 1995 report, the National Institute of Medicine conceded similarities between Gulf War syndrome and the side effects of pyridostigmine. It emphasized that past investigations showed the effects to be short lived. But it noted recommendations for studies of long-term effects, especially when exposure to pesticides and other chemicals is simultaneous, as occurred during the Gulf War.[56]

In March 1995, President Clinton announced the formation of a presidential advisory committee on Gulf War syndrome. Providing the committee with $13 million to oversee research, the president promised to "leave no stone unturned" to find the causes of the problem.[57] Months later, a study showed that a combination of pyridostigmine and insecticides in animals caused the kind of nerve damage the veterans had reported.[58] As related to human beings, however, the matter remained unsettled.

In August 1995, the Pentagon issued a health evaluation of some 10,000 of the 50,000 veterans who suspected their illnesses to be related to the Gulf War. Only 5 percent claimed they had been exposed to nerve agents, but 70 percent said they had taken pyridostigmine. The Pentagon study found no evidence for a "new or unique illness or syndrome." But it promised that the Defense Department would "maintain an ongoing search for unique symptom/illness patterns."[59] At the same time, the president's advisory commission announced its determination to pursue the matter and to issue findings by the end of 1996.[60]

Media reports continued to keep the issue in public view. At the end of 1995, *Life* magazine ran a cover story about children with birth defects born to parents of Gulf veterans. Laced with pictures of disfigured children, the article was titled "The Tiny Victims of Desert Storm." The provocative title did not match the uncertainties presented in the text.

Whether the children's problems were related to their parents' Gulf experience remained unclear. Robert Roswell, executive director of the government's Persian Gulf Veterans Coordinating Board, claimed in the article that a causal link between Gulf service and subsequent illness had not been established. But neither could he dismiss the possibility. "The one argument that does deserve further study [concerns] the combination of pyridostigmine bromide with pesticides."[61]

Diana Zuckerman directed the 1994 Senate Veterans' Affairs Committee report. She believes that research evidence pointed to pyridostigmine, either alone or in combination with pesticides, as the most likely cause of at least some cases of Gulf War syndrome.[62] Whether alone or in combination with other chemicals, the drug deserves careful investigation as a possible contributor.

Iraqi Arsenals Destroyed?

The third unresolved issue in the aftermath of the Gulf War concerns uncertainties about Iraq's biological and chemical warfare programs. The wisdom of allowing Saddam to remain in power has become a matter of debate. But in the wake of victory, the coalition was confident that Iraqi military power would be checked.

On April 3, 1991, the Security Council passed Resolution 687, which imposed a variety of constraints on Iraq. More than any other council member, the United Kingdom insisted that Iraq's chemical and biological weapons be destroyed. Some countries, notably France, expressed skepticism that compliance could be verified. Threatening a veto if the resolution failed to provide for total elimination, the British prevailed.[63]

The resolution required Iraq to reaffirm its obligations under the 1925 Geneva Protocol and to ratify the 1972 Biological Weapons Convention. Further:

> Iraq shall unconditionally accept the destruction, removal, or rendering harmless, under international supervision, of . . . all chemical and biological weapons and all stocks of agents and all related subsystems and components and all research, development, support and manufacturing facilities.

Resolution 687 instructed Iraq to inform the UN secretary general of the locations, amounts, and types of these items. He was to appoint a special commission (UNSCOM) to "carry out immediate on-site inspection of Iraq's biological, chemical and missile capabilities, based on Iraq's declarations and the designation of any additional locations by the Special Commission itself." UNSCOM was empowered to destroy, remove, or render harmless prohibited items or facilities.[64]

On April 7, Iraq grudgingly agreed to the United Nation's terms. "The resolution's clauses," complained the Iraqi foreign minister, "hold Iraq and its people hostage to the will of a number of superpowers so that they can control its resources and requirements of food, clothing and the modern life." Calling the requirements unjust and unfair, he concluded nevertheless that Iraq had no alternative but to accept the resolution.[65]

IRAQI CHALLENGES TO UN INSPECTORS

Iraq's reluctant agreement foretold its sporadic willingness to cooperate. Efforts by UNSCOM to fulfill its mandate proved at once frustrating and encouraging. Security Council pronouncements underscored Iraqi efforts to deceive. But the council would not relent, and its continued pressure produced considerable compliance. The pattern of Iraqi obfuscation followed by disclosure was dubbed "cheat and retreat."[66]

In the first year of inspections, UN teams faced physical threats. A Security Council statement in June 1991 noted that Iraqi soldiers "fired small arms into the air" when a nuclear inspection team tried to photograph vehicles leaving a site.[67] Such intimidation drew repeated criticism from the Security Council.[68]

In October 1992, UN inspectors informed the council that Iraq had been conducting a "systematic campaign of harassment, acts of violence, vandalism to property and verbal denunciations and threats at all levels." This prompted the sternest UN reaction since the Gulf War. The Security Council president warned Iraq to cease these actions or face "serious consequences."[69] When, in January 1993, the Iraqis prevented inspectors from entering the country on non-Iraqi airplanes, Western powers launched military strikes. Iraq relented.[70]

Iraqi threats of physical harm subsequently diminished, but efforts to conceal and mislead continued. Kathleen Bailey, an arms control analyst, wrote about UNSCOM's first 29 inspections. Conducted between May 1991 and April 1992, they included six for chemical weapons, two for biologicals, and one covering both. The remainder were for nuclear or missile programs. Bailey so often repeats the same characterization for the inspections that she uses an abbreviation: "CC&D," camouflage, concealment, and deception.[71]

Despite Iraq's CC&D, Bailey conceded that UN successes "have encouraged arms control planners to press for more intrusive inspections" in other international agreements.[72] Similarly, while criticizing Iraqi behavior, the Security Council president reported progress.[73]

UNSCOM 11, the third chemical weapons inspection, took place during the first week of September 1991. One of the locations was Al Taji, where inspectors found 6000 undeclared nerve-agent containers for rockets. When asked for documentation about the containers, the Iraqis said it had been stolen, although the facility was under heavy guard.[74]

UNSCOM 17, the fifth chemical weapons mission, lasted from September 27 through November 14, 1991. Inspectors found

205 tons of mustard agent at Al Muthanna, 75 tons less than the Iraqis had declared. Iraq said the missing quantity was due to a paper error. The team also found traces of the nerve agent sarin. After first denying its presence, the Iraqis said the sarin must have been from past experiments.[75]

The first biological weapons inspection, UNSCOM 7, lasted from August 2 to August 8, 1991. The Iraqis told the inspection team that they had never made biological weapons. But at Salman Pak, the team discovered

> a capability to research, test, produce, and store biological agents usable as BW [biological weapons]. Evidence indicated that there had been fermentation, production, aerosol testing, and storage equipment on-site.[76]

When confronted with these findings, Iraqi officials admitted that they had previously had an offensive program, but for research only. Some inspectors suspected that biological weapons had been developed and moved elsewhere.[77]

Resolution 678 prohibited trade with Iraq, except for medicines or other items for humanitarian needs. Inability to sell oil barred Iraq from its major source of hard currency. A principal incentive for Iraqi cooperation was the promise of lifting sanctions. The resolution mandated a review of Iraq's behavior every 60 days. When compliance was confirmed, the sanctions would be removed.

By 1994, Iraq seemed more directed at having the sanctions removed. In a retrospective issued in April 1995, UNSCOM reported having received "cooperation from Iraq in setting up and now in operating" a monitoring system. At the same time, the commission complained of Iraq's "refusal or inability to produce the documentation relating to its past programmes and [its] frequently changing accounts of certain elements of its programmes."[78] The finding prompted the Security Council again to demur from lifting sanctions. UNSCOM's monitoring and verification regimes could account for Iraq's behavior in some respects but not all.

The commission's April 1995 report reviewed Iraqi compliance in four areas: missile, nuclear, chemical, and biological. Despite contradictory information from the Iraqis, the 37-page document said that most areas were being adequately monitored. The exception was Iraq's biological warfare program.

According to the report, Iraq insisted that it "had no biological weapons-related activities." Rather its military program had been only in basic research, which it discontinued in 1990.[79] Iraqi officials acknowledged importing large quantities of growth media in 1988, ostensibly "for the purposes of hospital diagnostic laboratories." The commission was skeptical:

> This importation of media by types, quantities and packaging is grossly out of proportion to Iraq's stated requirements for hospital use. Iraq explains the excessive quantities imported and the inappropriate size of the packaging as being a one-of-a-kind mistake and attempts to justify the import as appropriate and required for medical diagnostic purposes.[80]

Iraq declared that from 1987 through 1994, hospital consumption of media had amounted to less than 440 pounds a year. But in 1988 alone Iraq imported 39 tons. The commission noted that the media deteriorate rapidly when removed from containers. Therefore, packages for hospital procedures normally hold a few ounces. The 1988 imports to Iraq came in large drums, each containing as much as 200 pounds.

> This style of packaging is consistent with the large-scale usage of media associated with the production of biological agents. The types of media imported are suitable for the production of anthrax and botulinum, known biological warfare agents researched by Iraq in its declared biological military programme.[81]

Other Iraqi biological activities raised more suspicions. The commission was able to account for only 22 of the 39 tons of media that Iraq imported in 1988. Iraq claimed that the missing 17 tons had been distributed to hospitals in 1989 but had been destroyed along with documentation about storage and consumption.

Iraqi officials presented documents to prove that the media had been received at a Ministry of Health storage site. It later admitted that the documents had been "recreated" and now claimed that "all originals have been destroyed, misplaced or lost." Meanwhile, the commission obtained information that Iraq purchased media in 1989 and 1990, which also were delivered in large packages. "This undermines Iraq's explanation that the [1988 purchases] were a one-of-a-kind mistake."[82]

Similar skepticism shadowed other behavior regarding materials that could be used to develop biological weapons. For example, the commission identified virulent anthrax strains that it said Iraq tried to order. "Iraq flatly denies this, despite confirmation to the Commission by the potential supplier."[83]

UNSCOM's Tempered Approach

In spite of obstacles and lies, the UN commission doggedly pursued its work. Monitoring biological activities is difficult in any case, and Iraq's behavior made the problem harder. But thus far, with few exceptions, Iraq has not physically prevented the UN team from instituting a monitoring regime. Accordingly, the commission has made a special effort to maintain an inventory of dual-purpose items—materials for nonmilitary purposes but also useful for a weapons program. UNSCOM's monitoring "modalities" are expansive. They include

> on-site inspections (with or without prior notice); aerial surveillance; interviews with key personnel at monitored sites; examination of site records; updating of inventories; continuous flow monitoring and sensor-activated camera monitoring; sample taking; notifications of transfers within Iraq of inventoried items; and notification of modification, import or other acquisition of dual-purpose biological research and production equipment of dual-use character.[84]

The commission has identified some 80 sites where it has tried to establish these verification protocols. They range from laboratories in hospitals, universities, and the food industry to drug-producing facilities, breweries, and distilleries. Presumption that work at these locations has been legitimate is tempered by Iraq's history of obfuscation. UNSCOM's conclusion:

> [T]he failure of Iraq to disclose fully all aspects of its past biological military research programme means that the

Commission cannot be certain that its monitoring programme in the biological area is covering all the sites, facilities and capabilities that require monitoring under the plan approved by the Security Council.[85]

The findings in UNSCOM's formal reports were reflected in observations by individual inspectors. At the annual meeting of the American Society for Microbiology in May 1995, six inspectors discussed their experiences. Each concentrated on a different aspect of the biological verification process—the utility of declarations, on-site inspections, aerial surveillance, problems posed by dual-purpose equipment, sampling measures, and remote monitoring techniques.[86]

All the inspectors felt their experiences provided lessons about biological arms control. But their emphases varied about the ability to detect Iraqi activities.

For Anna Johnson-Winegar, requiring Iraq to declare locations, equipment, and supplies showed that declarations "can be a very useful tool in analyzing a country's capability for biological warfare."[87] Others were more skeptical. David Huxsoll worried about equipment for vaccine and biopesticide production that could also be used in biological weapons programs.[88] Raymond Zilinskas felt stymied about facing Iraqi "collective amnesia."[89]

In a larger sense, the UNSCOM experience has been invaluable. It has served as a pilot study in the area of monitoring arms agreements. No verification protocol will provide for unlimited access, as has been the case with Iraq. But the UNSCOM experience has enabled assessments of techniques that will be useful even in a limited inspection regime.

Iraq's obduracy has been a favor to the international community. If Iraqi officials had been forthcoming and inspections made easy, the process might have seemed deceptively simple. UNSCOM would not have had to develop techniques to overcome Iraqi intransigence.

To its credit, the special commission has not flinched from criticizing Iraqi obfuscation and devising approaches to circumvent it. Although the commission acknowledges that information was incomplete, UNSCOM's tenacity has filled many gaps. As long as access to sites is permitted, a monitoring system can help deter development of prohibited programs:

Experience has shown that, even when initially presented with inadequate declarations, the Commission has been able, through the deployment of its various resources and the exercise of its inspection rights, to elicit the information required for the system to be established.[90]

This is not to minimize the importance of missing information, especially in the area of the biological growth media. Saddam may well have hidden material. But establishing that the material is missing is itself important. And the inspectors remain unrelenting. "Confession is not enough," said Rolf Ekeus, head of UNSCOM, about the missing culture material. "We need to get our hands on it and destroy it."[91]

Two months after UNSCOM's April 1995 report, Ekeus found that even more material was missing than the 17 tons acknowledged earlier. In response to Ekeus's charge, Tariq Aziz, Iraq's deputy prime minister, promised to provide more details about the biological program if the United Nations closed its files on Iraq's other weapons programs.

Members of the Security Council reacted in fury. Britain's representative, David Hannay, rejected the notion of a deal. "We're not buying carpets," he said. "We're talking about compliance with Security Council resolutions." Similarly, U.S. representative Madeleine Albright castigated the Iraqis for acting "as if they were in a bazaar." She dismissed the idea that they can decide "what they will comply with and not comply with."[92]

Sanctions by the United Nations and persistence by UNSCOM were not making life easy for the Iraqis. In July 1995, an Iraqi scientist contradicted earlier denials that Iraq had developed biological weapons. Rihab Taha, a microbiologist whom the United States suspected of heading the program, admitted to UN monitors that Iraq had produced 5300 gallons of *Clostridium botulinum* and 158 gallons of *Bacillus anthracis*. Enough to kill millions of people. Iraqi officials said the stocks of organisms were destroyed before the Gulf War. Now they expected the Security Council to end the trade embargo. United Nations inspectors said that analysis of the new data would take "some length of time." Ambassador Madeleine Albright remained cautious:

Obviously, it helps when they come up with information to clear their record. On the other hand, they shot themselves in the foot when they denied they had all this. It makes everybody doubt the veracity of what they say. It makes logical people wonder what else there is.[93]

Then, in August, two of Saddam's sons-in-law (who were brothers) and their wives defected to Jordan. One of them, Hussein Kamel Hassan, had been in charge of developing Iraq's weapons of mass destruction. His defection prompted Iraq to admit a larger program than previously acknowledged, including bombs and scud missiles armed with biological agents. Rolf Ekeus welcomed "the new language in Baghdad." But he reserved judgment about whether the weapons had been destroyed, as Iraq maintained, and whether Iraq was now entirely forthcoming. In December, he reported to the UN Security Council that Iraq was still withholding important documents about the program.[94]

Sustained pressure has moved Iraq to acknowledge aspects of its weapons programs that were previously concealed. The UN experience with Iraq has thus been a partial success.

The word "success" should be understood in a particular context. The need to inspect Iraq in the first place, like the other two unresolved issues reviewed in this chapter, represents a larger failure. They are all sad legacies of the 1980s. Had Saddam Hussein's chemical attacks against Iran been seriously protested in 1984, chemical and biological threats from Iraq in the 1990s would have been unlikely.

There would have been no worries about Gulf War veterans suffering from the effects of these agents. Or from pyridostigmine bromide or the anthrax vaccine. Nor would people later be wondering why Iraq failed to use its chemical and biological weapons during the Gulf War. Indeed, an earlier strong response by the international community might well have dampened Saddam's thoughts about invading Kuwait.

True, the Gulf War has provided an unusual opportunity to develop verification and monitoring procedures. The cause of biological and chemical disarmament in that sense has benefited. But the chance to inspect was hardly worth the suffering and death from Iraqi aggression.

Beyond the unanswered questions discussed in this chapter, the Gulf War stands as a divide on other biological and chemical warfare issues. Subsequent incidents have heightened concerns about poisons as terrorist weapons. Also in the wake of the war, international efforts to eliminate these weapons from the planet have intensified. Part III deals with these issues. As in the book's first two parts, its underlying theme is the conflict between morality and perceived interest. Despite the generally acknowledged immorality of these weapons, some persons and nations see having them—even using them—as in their interest. The challenge is to bury that disposition. Before reviewing current efforts to eliminate biological and chemical weapons, we turn to a particularly troubling by-product of their availability: their use as terrorist weapons.

New Challenges

Tools of Terror

As Kyle Olson flew over the Pacific Ocean, he wrestled with conflicting feelings. He was on his way to Tokyo where three days earlier, on March 20, 1995, sarin nerve gas had been released in the subway system. Olson had predicted the likelihood of such an event. Although his prescience brought a sense of satisfaction, he wished he had been wrong. The attack killed 12 passengers and injured 5500.

Boyish-looking, the 40-year-old Olson was executive vice president of the Chemical and Biological Arms Control Institute, a think tank in Alexandria, Virginia. His flight meant interrupting a whirlwind of interviews and television appearances since the Tokyo attack. He and institute president Michael Moodie had been besieged with inquiries about the event's implications for the United States. Olson wanted to gain a full understanding by talking to Japanese officials. A visit the previous December concerning the city of Matsumoto had prompted his prediction in the first place.

From Matsumoto to Tokyo

The Tokyo incident created headlines throughout the world. But it was not the first sarin release in a Japanese city. Nine months earlier, an event in Matsumoto was a forerunner. Located 100 miles west of Tokyo, Matsumoto is a mountain resort with a population of 200,000. Soon after 10:30 P.M. on June 27, 1994, people in a residential area began to feel ill. Some said their symptoms started with an odd feeling in the eyes and throat. Many complained of nausea, vomiting, and difficulty breathing. "It was like wearing very strong glasses—shapes became distorted," recalled one victim. "I started hearing things, had nausea, and the light broke up into several colors."[1]

Rescue workers found scores of unconscious victims and others who were nauseated and having difficulty breathing. By the next day, seven people had died and more than 200 were ill. Fish in a nearby pond were dead, and birds, dogs, and other animals lay dead in the street. Witnesses said that during the night a strange fog had descended on the town. Authorities were puzzled because of the absence of chemical plants in the area.[2]

Although the media outside Japan paid little attention after the initial report, Japanese newspapers continued to give the incident broad coverage. On June 30, the *Mainichi Daily News* reported that doctors found that victims had dramatically lower levels of acetylcholinesterase. The enzyme is necessary to maintain normal nervous system activity. The findings suggested exposure to organophosphates, which reduces levels of the enzyme.

Organophosphates are ingredients in some insecticides. The *Mainichi* article said police suspected that a neighborhood man was mixing chemicals to make a pesticide for his garden.[3] Organophosphates are also components of nerve gas, but this was not mentioned in the article.

Two days later, the suspect was identified as a farm equipment salesman who had been the first person to call for an ambulance. From his hospital bed he denied the police allegations. The *Mainichi Daily News* said: "If the unnamed man is telling the truth, [the incident] may well have been an accident." The story was headlined "The Poison Gas Mystery."[4]

On July 4, the gas was identified as sarin. But its synthesis was being portrayed as an accident. The sarin identification was based on analysis of water in a pond at the accused man's home. A Japanese historian of science suggested that the problem was based on coincidence: "It's quite plausible that, although unstable, the gas was given off by ingredients discarded in the pond."[5]

Japanese authorities continued to investigate while holding that the sarin could have resulted from an accidental concoction. At the same time, investigators confessed their ignorance about the agent. They reported making inquiries overseas to learn more about it.[6]

Two months after the incident, Japanese authorities remained baffled, and speculation abounded. Guesses about the cause ranged from leaky poison-gas shells left over from World War II to an attack by the North Korean government.[7]

Kyle Olson's visit in December 1994 left him with a different impression from the one presented by the authorities. He allowed that Japanese intelligence and counterterrorism investigations may have been taking place out of public view. But in his postvisit report he wrote:

> The Matsumoto incident has generally been referred to by authorities and the media in Japan as "the accident." There is compelling evidence that whatever the complete story of that deadly June night turns out to be, the events in that quiet city were anything but accidental. This case deserves further attention as the potential harbinger of the next phase of terrorist horror.[8]

Olson thought the incident was premeditated for three reasons. First, although sarin is not technically difficult for a chemist to make, neither is production a trivial exercise. Synthesis is "simply not something that can be done by accident." Second, although normally rainy in June, several dry days had been predicted before the time of the incident. In dry weather the sarin was less likely to be washed away. The forecast was accurate. "Someone," Olson theorized, "anticipated a break in the weather and took advantage of it." Finally, the absence of a container to transport or generate the nerve gas "indicates a planned effort to conceal the identity" of the responsible party.[9]

The fact that no one claimed responsibility for the release also suggested to Olson that the incident was an experiment. In consequence, his report said, "the person(s) responsible for Matsumoto certainly understand now that a significant quantity of nerve gas, delivered into a warm, crowded urban site (such as a Ginza department store, or major subway system) could have catastrophic consequences."[10]

Japanese authorities would say only that they were investigating many possibilities. But a news report on March 19, 1995, nine months after the incident, indicated that they were now persuaded the Matsumoto attack was intentional. Moreover, they believed it was a trial run by terrorists.[11]

The timing of the story was remarkable. The following day, a nerve agent was released in the Tokyo subway system. The police quickly identified the agent as sarin. The Matsumoto experience had indeed proved to be a preparation for Tokyo.

The Tokyo attack reverberated everywhere. The next day's front page of the New York *Daily News* reminded readers of their own vulnerability: "The nerve gas attack on Tokyo's subway system laid bare an awful truth in New York: The city is not prepared for chemical terrorism."[12] When an Olympus Airways flight from Athens landed at Kennedy airport, the 200 passengers could not disembark for eight hours. Poison gas was reportedly on board, and the plane had to be searched. None was found, and the charge was deemed a hoax.[13]

The Hackensack, New Jersey, *Record* announced that police patrolling the subway line between Hoboken and New York City "were on the lookout for unusual packages." Underground employees "were ordered to be on the alert for anything suspicious."[14] Other cities also sought to allay the fears of passengers. The London Underground announced that "long practiced anti-terrorist measures" were in place. Moscow commuters were assured that "safe rooms" had been set up to defuse unexploded bombs.[15]

Even before the Tokyo event, Japanese investigators had become suspicious that a religious sect called Aum Shinrikyo ("Supreme Truth") might have been involved with the Matsumoto incident. A few days after the Tokyo attack, police searched the Aum compound near Mt. Fuji. They discovered tons of chemicals needed to make sarin and other poisons.

In subsequent weeks, leaders of the 10,000-member sect were arrested, including its charismatic head, Shoko Asahara. Several Aum members held advanced degrees in science. They had built chemical plants, in which sarin and other agents were produced, within the sect's building complex. In police custody, they confessed to using the material in Matsumoto and Tokyo.

The sect's leaders said that, for the subway attack, sarin was packed into 11 sealed plastic bags. Aum members placed the bags in five subway cars at rush hour and then pierced them with sharp umbrella tips.[16] A U.S. Senate committee staff investigation yielded a breathtaking observation:

> It was only a fortunate mistake by the Aum in the
> preparation of the special batch of sarin used that day and
> the inferior dissemination system to deploy it that limited
> the number of casualties. If not for these mistakes, the Staff
> has been told by chemical weapons experts, tens of
> thousands could have easily been killed in this busy subway
> system that moves over five million passengers a day.[17]

In searching the Aum compound, the police also found evidence that the sect was developing biological weapons. A large amount of *Clostridium botulinum* was discovered along with 160 barrels of media used to grow the bacteria.[18] The bacteria produce a toxin that is lethal in tiny amounts and has been stocked by nations with biological warfare arsenals. Other biological agents were being produced in a special building, which Japanese authorities sealed. At the end of 1995, police had still not entered, apparently because of concern about the danger of its contents.[19]

The motive for the Tokyo attack was bizarre. The police found documents in Aum files predicting that gas attacks would contribute to killing 90 percent of the people in major cities and that the world would undergo a cataclysm by 1997. Only the truly faithful (to Aum) could survive.[20] The Tokyo incident was an experiment toward fulfilling the prophecy. Most of the world's population would be annihilated, the group's leaders apparently believed, in a bath of chemical and biological warfare that Aum would initiate.

In the weeks following the Tokyo attack, other suspicious chemical incidents occurred. On April 19, strange-smelling fumes were detected in the main railroad station in Yokohama. More than 500 people went to hospitals with sore throats, coughs, and other symptoms. Two days later, fumes in a Yokohama department store sent 25 people to the hospital with similar problems.[21]

On May 5, a cleaning woman found two plastic bags in a rest room at a busy Tokyo subway station. Police later determined that Aum agents had placed the bags there, one filled with sodium cyanide, the other sulfuric acid. They were rigged with an automatic trigger to combine the chemicals, which would produce cyanide gas. The device was near a vent with a fan that pipes to a subway platform. Had the cleaning woman not intervened, thousands of people could have been killed.[22]

The police were unable to establish whether Aum Shinrikyo was responsible for all the incidents or whether some were "copycat" efforts by others. In any case, a previously uncommon form of terrorism now seemed to be taking root.

For Americans, the events in Japan were overshadowed by the bombing of the federal building in Oklahoma City on April 19. Explosives hidden in a truck killed 169 people, more than in any previous terrorist incident on American soil. Moreover, ongoing efforts to rescue trapped victims were shown on national television for weeks afterward. The cratered building became a consuming public image.

United States authorities believed the bomb perpetrators were right-wing militants who considered the government illegitimate. The media and public officials began to focus on extremist organizations, concerned by their potential for further terrorist action.

Oklahoma City highlighted terrorism by explosives. But a month after the bombing, news reports showed that biological and chemical terrorism had not been forgotten. Ohio police indicted a laboratory technician named Larry W. Harris on charges of receiving stolen property and obtaining bacteria by fraud. As mentioned in Chapter 1, Harris acquired three vials of *Yersinia pestis,* the bacteria that cause bubonic plague. On May 5, he ordered the organisms from the American Type Culture Collection (ATCC), a biomedical supply firm in Rockville, Maryland. He mailed the order under the letterhead of a fictitious research laboratory.[23]

The bacteria had not arrived by May 9 as expected, and Harris telephoned ATCC expressing impatience. The call prompted concern by the firm's officials. They contacted the Centers for Disease Control and Prevention (CDC) and other government agencies, which then became curious. In answer to a CDC official, Harris said he wanted the bacteria to conduct "biomedical research using rats to counteract imminent invasion from Iraq of supergerm-carrying rats."[24]

Several of Harris's work colleagues indicated that he openly spoke of being a white supremacist. A police search of his home found racist and anti-Semitic literature along with his membership certificate in the white supremacist organization Aryan Nations. The county prosecutor said that Harris's views were irrelevant to the case. But state civil rights organizations were alarmed. A spokesman for the Anti-Defamation League of B'nai B'rith believed "he had more on his mind than an antidote to Saddam Hussein."[25]

In July, U.S. authorities charged Harris with engaging in mail and wire fraud. Just before a scheduled federal court trial in November, Harris pled guilty to wire fraud in exchange for a maximum six-month jail sentence.[26]

Had Harris not made a follow-up telephone call to ATCC, he would have raised no suspicions. The bacteria were already en route, and his shipment was like thousands by the firm to laboratories throughout the world. Pathogens such as *Yersinia pestis* might be wanted, for example, to develop vaccines or to investigate their life cycles. Virtually all are used for legitimate purposes—research, hospital sterilization checks, commercial processes. But the potential for misuse is clear.[27]

Making Biological or Chemical Weapons

In 1989, Senator John Glenn observed that a facility the size of a large room "could turn out very substantial amounts, tons even, of nerve gas or biological weapons." Glenn exaggerated. Not about the quantity of weapons, but about the size of the room. To make these weapons, a small kitchen would do.

Of course, stockpiling the agents would require more space. And if delivery systems such as grenades and bombs are considered, the matter becomes more complicated. But production alone can be done in a small area. An ounce of biological agent in a half-gallon of growth medium, Glenn noted, could produce enough material to "sicken or kill perhaps as much as 95-percent of a population the size of Washington, D.C."[28]

What the senator learned in 1989 about the ease of producing chemical and biological weapons had been long understood by knowledgeable people. Processes to make mustard gas date to the mid-nineteenth century. The cause of thousands of blistering and respiratory casualties in World War I, mustard agents are technically easy to make.

One method consists of combining hydrochloric acid and thiodiglycol. Hydrochloric acid is available in huge amounts for industrial purposes. Thiodiglycol, which is used in the manufacture of dyes and inks, is a product of two chemicals commonly found in science and medical laboratories: ethylene oxide and hydrogen sulfide.

The manufacture of nerve agents is more complicated. But supplied with precursor chemicals, a graduate student in chemistry should have little difficulty synthesizing them. Like mustard gas, sarin and other chemical agents can be produced in a variety of ways.

Although the nerve agents tabun, sarin, and soman were developed in Germany in the 1930s, the allied powers did not learn about them until the final year of World War II. Since the late 1940s, techniques for making the agents have appeared in the open scientific literature. More recently, nontechnical publications have listed the materials necessary for production, if not the conditions required to make them interact.[29]

A typical route to making sarin begins with raw materials that are familiar to any chemist. Edward Naidus is a retired technical director at American Cyanamid. He frequently worked with sarin's precursor chemicals during his 50-year career in academia, government, and industry. They are used for a variety of purposes in many laboratories, he said. "You could probably get them through a supply-house catalogue. You'd just need to write under a company or university letter head."[30] Two of the materials, isopropyl alcohol and methyl alcohol, can be purchased from a local drugstore.

Naidus's comments echoed those of an army chemical weapons analyst who testified before Senator Glenn's committee in 1989. He was asked about the difficulty of producing chemical weapons. The corrosive nature of some chemicals might require platinum or gold-lined vessels, he said, but "you can go to any first year organic chemistry book and at least get the basic chemistry for the production of many of these agents."[31]

Synthesizing sarin would be dangerous. Its appeal as a weapon derives from its remarkable potency. A single drop can kill within minutes of exposure to the skin or by inhalation of the vapor. Moreover, clothing offers scant protection, because the agent can penetrate ordinary outerwear.

Large-scale production of sarin therefore is done in a closed system. But what if someone wanted to produce a pound or two at home? For a chemist, that would not be difficult, knowledgeable people agree.

The vessels for mixing the chemicals would have to be corrosive-resistant—Pyrex glass would do. A venting system for drawing toxic fumes out of the work area would be important. And to insure survival, the operator could wear protective gear, much like a soldier dressed for gas war. Information about when to add each chemical, and at what temperature, has been in the open literature for nearly 50 years.

To make a biological weapon would be even easier. Methods for culturing bacteria have been known for a century, and refinements have enhanced techniques over the years. A publication in the 1960s, for example, reported that adding trypsin to the growth medium for *Clostridium botulinum* increased the potency of the toxin as much as 400 times.[32]

Basically, you need seed bacteria, said Nancy Connell, a microbiologist at the University of Medicine and Dentistry in New Jersey.[33] An organism such as *Bacillus anthracis*, the cause of anthrax, can be obtained from supply firms such as ATCC. Supply firms also provide culture media for enhancing bacterial reproduction. But bacteria can grow in less specific media as well.

Could *Bacillus anthracis*, long considered a likely biological weapon, be grown at home? "I wouldn't be surprised," Connell answered. Foods that contain amino acids and other nutrients might be effective media. "Any source of protein would do." The at-home media would first have to be autoclaved to kill other organisms that would compete with the anthrax bacilli.

Bacteria can divide every 20 minutes. Thus one bug can become a billion in ten hours, and a few would yield a formidable arsenal in a matter of days. Deprived of media, certain bacteria, such as anthrax, will sporulate. That is, they develop a hard coating in which they can remain dormant for years.

If the spores reach a warm, moist environment, such as human lungs, they revert to active vegetative states. They then become devastatingly infective. Inhaling a few thousand anthrax spores, less than a pinhead in size, can kill someone not quickly treated with antibiotics.

Making bioweapons in a kitchen would be less safe than in a laboratory where air is drawn from the work space. But wearing a protective mask could help prevent an operator from inhaling the organisms.

How could biological or chemical agents be disseminated? The possibilities are limitless. Past army tests included spraying San Francisco with bacteria from a boat offshore, tossing light bulbs filled with bacteria on New York City subway tracks, and spreading chemicals with fans from perforated suitcases in a Washington, D.C., bus station. Testers sprayed chemicals from the rear of slowly moving cars in St. Louis and from rooftops in Minneapolis. As described in Chapter 2, a low-flying airplane blanketed several states on one trip.

If a city's water purification system were blocked, a gallon of botulinum toxin in the water system could theoretically kill millions. An innocent-looking fishing boat might circle Manhattan island, blowing anthrax spores from an inconspicuous aerosol generator. Again, casualties could be in the millions. The boat and all traces of its activities would long be gone before the massive outbreak of infections was apparent.[34]

Kyle Olson, formerly of the Chemical and Biological Arms Control Institute, is convinced that, whatever the manner, terrorists will use biological and chemical agents. "It is not a question of if; it is a question of when. It is also a question of what magnitude it would be."[35] Olson's prediction of the Tokyo nerve-gas attack adds chilling credibility to his observation.

Methods for developing gas and germ weapons have long been understood. The number of chemists and biologists has grown sharply since World War II. Millions of Americans have had substantial exposure to college-level biology or chemistry.

Terrorist experts Robert Kupperman and David Smith write that biological weapons were not used in World Wars I and II because of their "moral abhorrence." They believe, however, that by the end of World War II the barrier had almost disappeared. Although giving no evidence, they contend that if the war had continued, "biological weapons would undoubtedly have been used" by the United States.[36]

The war ended 50 years ago. If the moral barrier had fallen by 1945 as Kupperman and Smith suggest, why have these weapons not generally been used?

In fact, a few terrorist organizations did try to acquire biological agents in recent decades. In 1972, a group in the United States called the Order of the Rising Sun was found with about 40 kg of typhoid bacteria cultures. In 1980, the German Baader-Meinhof gang was discovered with a culture of *Clostridium botulinum* in a home biological laboratory. And in 1984, the Rajneesh cult in Oregon contaminated salad bars in local restaurants with *Salmonella typhemurium,* the cause of typhoid. But as a government-sponsored assessment says, preparation of biological or chemical agents for use by terrorists has been rare.[37]

The rarity is all the more striking when considering the ease with which chemical and biological weapons can be developed. The answer seems related to the same moral inhibitions that have made their use rare among nations.

Behind Terrorism

Defining terrorism is elusive. The U.S. Departments of State and Defense see it as "premeditated, politically motivated violence perpetrated against a noncombatant target by subnational or clandestine state agents, usually intended to influence an audience."[38] The Office of Technology Assessment (an advisory body to the Congress that was eliminated in 1995) offered an ungainly, if more comprehensive, definition:

> The deliberate employment of violence or the threat of violence by sovereign states or subnational groups, possibly assisted by sovereign states, to attain strategic or political objectives by acts in violation of law intended to create a climate of fear in a target population larger than the civilian or military victims attacked or threatened.[39]

Difficulty in defining the term lies in the variety of intentions, values, and techniques that it encompasses. The many faces of terrorism were displayed in 1995. In the Middle East, Arab militants murdered Israelis and Americans in suicide car-bomb attacks. In the United States, a man was killed as he opened a letter bomb, the most recent victim of a 17-year vendetta by an elusive "Unabomber." The destruction of the federal building in Oklahoma City was done with a massive truck bomb. And the Tokyo attack was with nerve gas.

The purported motives of each incident were as disparate as their methods: to stop the peace process between Israel and its neighbors; to punish people committed to advancing science and technology; to demonstrate opposition to the U.S. government; to rehearse for larger attacks in anticipation of an Armageddon.

Inquiries into the sources of terrorism yield a range of shallow generalizations. Before the dissolution of the Soviet Union in 1991, some believed that terrorism was largely prompted by a Soviet-sponsored network. Others challenged the notion of a single network, holding that permissive behavior in democratic societies was responsible. One critic of both approaches blamed terrorism on the loss of connectedness between young people and their society. The answer was to "reconnect politicized young adults by involving them in mass-based movements for change."[40]

The weakness of these explanations is obvious. The Soviet Union has disappeared, yet terrorism persists. Some democratic "permissive societies" experience more terrorism than others. Moreover, terrorism is not unusual in less democratic countries—

Algeria, Egypt, Peru, and Colombia for example. Reconnecting disaffected young people sounds more like tautology than a pragmatic action plan.

Scholars have looked in vain for a "terrorist personality." A survey of the issue by law professor Richard Rubenstein led him to conclude:

> People who declare war on the state may be quite sane or seriously disturbed. They may be expressing repressed hatred of their parents or they may not. They may harbor intense suicidal impulses or may have decided, like good soldiers, to take unusual risks for a cause they believe in.[41]

Terrorists disproportionately are young men from middle-class or upwardly mobile working-class backgrounds. But they seem to have little else in common beyond the desire to shock, inflict pain, and intimidate. In asking why terrorists have tended not to use chemical or biological weapons, however, other criteria appear useful. David Long, who coordinated counter-terrorism policy for the U.S. State Department, noted that terrorists are highly influenced by group psychology. The groups, unlike individuals, often have their own terrorist personality.[42]

Brian Jenkins, another expert on terrorism, observed:

> Each terrorist group has its own repertoire, its own style of operations, its own *modus operandi*. The Irish Republican Army does not engage in the hijacking of airlines or kidnapping. The Italian Red Brigades kidnap and shoot journalists and others in the legs. West German terrorists seem to be planners.[43]

In addition to group characteristics, there appears to be a temporal style to terrorist methods, especially those in the Middle East. Hijacking airplanes was a preferred technique in the 1960s and 1970s. Blowing airplanes up in flight became more common in the 1980s. Car bombs detonated by remote control also were fashionable in the 1980s. More recently, suicide bombers and grenade throwers have become typical.

Some of these approaches become less inviting as preventive methods improve. Enhanced airport security, for example, has reduced the likelihood of skyjackings or midair bombings.

Moreover, a terrorist group may not confine itself to one method during a particular period. Even while sponsoring suicide attacks against Israelis, the Islamic extremist group Hamas continued knifings and kidnappings.

Nevertheless, just as a serial killer tends to repeat his pattern of murder, so do terrorist groups leave signatures on their methods. David Long argues that some groups may be happy about this. Associating one form of attack with a group may add to its ability to intimidate. But he also notes the importance of habit. If groups "have an expert in explosives or some other device, for example, they will continue to use his services, and he will teach others his methods, perpetuating the use of the tactic."[44]

Long's logic is less convincing when he turns to the subject of biological and chemical agents.

> The danger that a terrorist group might acquire a nuclear device or even biological or chemical agents is highly unlikely. . . . Stealing a nuclear device or chemical or biological agents would require breaching the elaborate security measures installed to prevent such an act. . . . Constructing a nuclear device from scratch would require fissionable materials, which themselves would have to be obtained. Chemical and biological agents would be relatively easier to obtain but would still require expertise to handle.[45]

Long's error is to lump nuclear, chemical, and biological weapons under one generalization. Although allowing that chemicals and biologicals would be "relatively easier" to obtain than nuclear devices, he implies that the difference is not great. It is therefore "highly unlikely" that terrorists will acquire any of them. Writing in 1990, Long did not have the benefit of the Aum Shinrikyo experience. But even before Aum, informed observers understood that making a chemical weapon was far less complicated than making a nuclear weapon. And a biological weapon, easier still.

A single person can produce lethal chemical or biological agents. But the number of skilled people required to manufacture a nuclear device is estimated to be at least six and as many as twenty. Moreover, the mix must include highly qualified scientists and engineers with the right specialty expertise.[46]

If producing a chemical or biological weapon is less difficult, why have they not been freely used by terrorist organizations? One of the reasons appears to be their unfamiliarity. Conventional weapons are well known and widely understood. Explosives are easy to obtain, experience with them has been vast, and they have been effective for terrorist purposes.

Easy as developing biological weapons would be, the technique requires some expertise. For a terrorist to establish a biological or chemical "signature" would require veering from old, familiar techniques. But there is another important reason that such weapons have not been favored by terrorists. As Livingstone and Douglass have observed, chemical and biological weapons "are really frightening and horrifying—even to terrorists."[47] Does this suggest a sense of moral inhibition? Most experts on terrorism place little weight on the value of moral constraint in considering their subject. The notion is rarely mentioned. An Office of Technology Assessment (OTA) report, for example, asked why biological weapons have not been widely used by terrorists. It listed six far-ranging answers: lack of familiarity; fear of alienating supporters by causing large numbers of fatalities; fear of an extreme response by the target country; fear of working with biological weapons; prohibition by a terrorist group's financial sponsors; awaiting someone else's successful use.[48]

The word morality never appeared on the OTA's list. But not all analysts have ignored the moral implications. Long writes that, apart from security measures, "the greatest constraints against nuclear, biological, or chemical terrorism rest with the terrorists themselves and relate to political expediency and political morality."[49]

The meaning of his remark is not entirely clear. Have terrorists been reluctant to use such weaponry because of their own sense of moral repugnance? Or have they refrained, as suggested by OTA, because *other* people's sensitivities would be offended and the terrorists' goals undermined? From Long: "As a media event, such a threat might indeed be gripping, but were it carried out, the catastrophe would totally overshadow the terrorist's cause and serve to escalate the alienation of the group's supporters even more."[50]

In either case, Long believes that terrorists are affected by the moral environment. Whether induced by internalized scruples or by the way a terrorist perceives an audience, a sense of morality has contributed to his selection of weapons.

Brian Jenkins is more direct. "Many terrorists consider indiscriminate violence to be immoral," he says. After years of studying the matter, Jenkins is

> convinced that the actions of even those we call terrorists
> are limited by self-imposed restraints that derive from
> moral considerations or political calculations. The growing
> volume of testimony from terrorists interviewed while still at
> large, interrogated in prison, or testifying at trials has, I
> believe, borne out that notion.[51]

Writing in 1985, soon after Iraq began to use chemical weapons, Jenkins offered another important insight. Terrorists appeared more interested in acquiring nuclear weapons than biologicals and chemicals. The reason related to his observation about morality:

> Terrorists imitate governments, and nuclear weapons are in
> the arsenals of the world's major powers. That makes them
> "legitimate." Chemical and biological weapons also may be
> found in the arsenals of many nations, but their use has
> been widely condemned by public opinion and proscribed
> by treaty, although in recent years the constraints against
> their use seem to be eroding.[52]

Others offer similar assessments: As more states take up biological weapons, the probability increases that terrorists will do the same. Indeed, Aum Shinrikyo leaders said they were inspired to develop chemical and biological weapons by publicity about Iraq's capabilities at the time of the Gulf War.[53] Thus, as the moral barrier against biological and chemical weapons falls among nations, the likelihood of use by subnational groups increases.

What can citizens do to protect against a biological or chemical weapons attack?

Is Defense Against Bioterrorism Possible?

Guarded confidence that the United States is prepared to respond to biological or chemical terrorism flows from many levels of government. A few days after the Tokyo subway attack, a spokesman for New York City's transit police was asked on national television how such an incident can be prevented. "Our officers are well trained," he said. "They recognize the danger signs of a suspicious package under a seat." No guarantees, but the police "are trained to deal with that in a way that will not risk their lives or jeopardize the public."[54]

The city's health commissioner, Margaret Hamburg, offered similar assurances that New York could counter an act of biological or chemical terrorism. She noted that the city had participated in a mock terrorist attack about a year earlier. The exercise was held in cooperation with federal and local agencies. Organized by the Federal Emergency Management Agency (FEMA), the test simulated an attack with anthrax bacilli.

The bacteria hypothetically contaminated indoor and outdoor locations. Local officials indicated that antidotes and antibiotics were available, and emergency workers were trained to wear protective gear and masks.[55]

At the federal level, response to a large-scale attack with biological or chemical weapons would require the services of the National Disaster Medical System (NDMS). The NDMS is a consortium of four federal agencies that attend to medical needs in case of disaster. Typically, they are called upon during natural disasters—floods, hurricanes, and earthquakes. Beside FEMA, the consortium includes the Departments of Health and Human Services, Defense, and Veterans' Affairs.[56]

A terrorist incident would draw the direct interest of the Federal Bureau of Investigation and other agencies specializing in intelligence and crime. But biological or chemical weapons add a health dimension to the matter. The nation's *Federal*

Response Plan makes no distinction between man-made and natural disasters. Thus the medical response to an accidental biological or chemical release would be no different from a response to a terrorist attack.

The Public Health Service and the Centers for Disease Control and Prevention (both agencies are subordinate to the Department of Health and Human Services) are assigned to lead in either a biological or a chemical emergency. The designated response action to biological hazards is

> Assist in assessing health and medical effects of exposure to biologic agents on the general population and on high-risk population groups; conduct field investigations, including collection and laboratory analysis of relevant samples; advise on protective actions related to direct human and animal exposure, and on indirect exposure through biologic agent contamination of food, drugs, water supplies, and other media; and provide technical assistance and consultations on medical treatment of victims injured by biologic agents.[57]

The plan later discusses coordination and logistics in general but contains no other specific references to biological hazards. How effective the guidelines would be during an attack can only be surmised. But not all people taking part express confidence about existing arrangements.

James Rabb, an emergency response coordinator at the CDC, described his agency's activities during floods in the country's South and Midwest. The CDC was concerned from a public health perspective with water quality, bacterial and chemical contamination, and communicable disease surveillance.

A biological or chemical "event," Rabb said, would be treated much the same. He believed, however, that the federal government may be developing an "annex to the plan for a chemical or biological event." Observing that the plan does not now "tailor into terrorism," he hoped that the CDC and the military would "integrate about who responds how."[58]

In their 1993 analysis, Kupperman and Smith said that the United States is "virtually unprepared" to cope with biological terrorism.[59] The authors urged that FEMA oversee an improved biological "protective posture" for the country. Among their suggestions:

> the United States should pursue an all-embracing concept that includes proactive intelligence collection, . . . the technological ability to destroy clouds of pathogens, countermeasures such as lasers or bleach-saturated "counterclouds," protective measures (civil defense and public education).[60]

In June 1995, three months after the Tokyo attack, Robert Kupperman's view was essentially unchanged. He believed there had been increased interest in vaccine development and laser identification of biological agents since he and Smith made their suggestions. But the overall state of U.S. preparation for a biological attack remains "in deplorable shape."[61]

Kupperman reemphasized his earlier proposals: The public should be informed about the likelihood of a biological attack, and the nation should be better prepared to deal with mass casualities. This would include maintaining large supplies of medications, he said. Antibiotics should be stockpiled to counter biological agents, and atropine along with high-pressure injectors should be available to counter nerve agents.

Should people store these materials at home, and should needles be distributed? Kupperman hesitated. "I don't think it would be a good idea to give needles and atropine during periods of calm."

However valid his concerns about terrorism, Kupperman's proposed countermeasures seem unfeasible. His plan would require a massive diversion of public attention and resources. Ability to identify an invisible cloud of pathogens is limited and seems likely to remain so. The varieties of natural biological agents, dangerous or not, are almost incalculable. With a capability to engineer new strains, the notion that any device will be able to select out pathogens is questionable. (The devices are further discussed in Chapter 10.)

In any case, where would such devices be placed? In every subway, bus, train, school, office building in the country? A biological attack can take place anywhere—indoors or out. Kupperman and Smith's "all embracing" defense implies that monitoring devices, if they can be developed, would be in place everywhere, and constantly activated.

If an alarm were to go off, should a "bleach-saturated 'countercloud'" be let loose? What would happen to the people being sprayed with bleach—subway commuters, for example, during rush hour? Panic? Stampede?

The call for civil defense measures to counter a biological attack are reminiscent of army proposals 30 years ago. In 1966, the army conducted a mock biological warfare attack in the New York subway system. Testers dropped light bulbs filled with *Bacillus subtilis* on ventilating grills at sidewalk level and on the tracks as trains entered the station.

Less harmful than actual warfare agents, the test organisms still posed risks. Confirming what should have been obvious without testing, the bacteria spread as trains whooshed in and out of stations. More than 1 million commuters were unwitting subjects in the test. The army report's conclusion: "A large portion of the working population in downtown New York City would be exposed to disease if one or more pathogenic agents were disseminated covertly in several subway lines at a period of peak traffic."[62]

On the basis of the test experience, the army suggested countermeasures that were no more practical than the more recent proposals. One measure was to train subway personnel "to look for signs of covert use of biological agents." Another suggested "enforcement of an ordinance against litter." Neither training of personnel nor litter control is likely to offer protection. Germs can be dumped into the system in any number of ways—by light bulbs, as the army showed, or by other innocuous-looking containers.

The remaining army proposals were equally unconvincing: increase the number of trainmen "at critical political periods"; collect air samples "at peak workday traffic periods"; vaccinate "key personnel" against potential biological agents.[63]

Determining a critical political period is as elusive as Kupperman's "period of calm." The Tokyo attack occurred during a calm, "noncritical" period. The proposals to collect air samples and vaccinate personnel are equally problematic. The army considers dozens of agents to be potential biological warfare threats. Testing the air several times a day for these agents throughout the subway system is plainly impractical. Similarly with vaccinations. For some agents there are no vaccines. For agents that are susceptible to inoculation, the numbers and variety of vaccines to cover every person would be formidable.

This is not to say that vaccines or other measures would necessarily be useless. A person who has been vaccinated against anthrax stands a greater chance of resisting the disease. But the fit between a warfare agent and a vaccine would have to be precise. Novel strains of anthrax or other organisms can nullify the effort. As Keith Yamamoto, a University of California biochemist, told a Senate committee, new methods in genetic engineering can "render existing vaccines useless."[64] A civil biological/chemical defense program is more likely to result in costly self-deception than in true protection.

An understandable zeal to avoid panic has prompted officials to offer more assurances about existing protection than appears warranted. Enhanced intelligence might thwart a prospective attacker. But most steps in the name of defensive action, as urged by Kupperman, Smith, and others, will yield only an illusion of protection.

How to respond? Intelligence that allows for interception before an attack is an obvious immediate answer. But the importance of the moral curtain also deserves consideration. The international norm that has generally placed biological and chemical weapons beyond the pale needs to be cultivated and reiterated. Not that moral attitudes can be generated on demand. They can be enhanced, however, by pragmatic and substantive actions. To this end, strengthening treaty arrangements to prohibit these weapons is an important step.

CHAPTER 9

Banning Weapons, Catching Cheaters

"To oppose lists is to oppose verification," Barbara Rosenberg said in 1994. She was referring to talks underway in Geneva to strengthen the Biological Weapons Convention.[1] A proposed verification regime was being considered that would list scores of possible biological agents.

Donald Mahley, head of the U.S. delegation there, disagreed. "You get in trouble with making a list of every supposed agent." He worried that some countries might be tempted to use agents not included on the list. But by 1995 the United States had moved to accepting "a short list of maybe seven examples."[2]

Rosenberg speaks on biological weapons issues for the Federation of American Scientists, an organization of 3000 scientists dedicated to arms control. She has seen the United States position move closer to her organization's, a position held as well by many other countries.

The United States prefers the term "transparency" rather than "verification" to describe the proposed regime. But the difference has become more one of nuance than substance. After earlier reluctance, the United States now accepts the idea of international inspections to assure compliance with the treaty. Not that everyone is happy about this. Pharmaceutical companies worry that outside inspectors might learn about their proprietary secrets. "We don't need to know about the genes that make their processes faster than somebody else's," Rosenberg says in response, just "if the host has pathogenic characteristics." Rosenberg is more confident than Mahley that such explanations will put the skeptical companies at ease. Whatever their differences, Rosenberg and Mahley share much common ground. They seek a stronger Biological Weapons Convention.

Not so for Kathleen Bailey, a former assistant director of the Arms Control and Disarmament Agency. She opposes lists, inspections, and indeed any verification protocol for either the Biological or the Chemical Weapons Conventions. She thinks the treaties are unverifiable.[3]

Bailey published two books in 1995. One of them, *The UN Inspections in Iraq,* was issued by an academic press. The other, a novel titled *Death for Cause,* is about a terrorist group that uses biological weapons.[4]

> "The lobby area is contaminated [with anthrax spores]."
>
> "How soon will we know if this is all there is?" the Prime Minister pressed.
>
> "Our technology does not allow us to just zoom in on a concentration of an agent. . . . What we don't know is whether there have been attacks outside of the center, although there is no indication of this. We don't know if there are cannisters that have not yet popped open."

The novel demonstrates the improbability that a civilian population can be protected against a biological attack. Nevertheless, Bailey supports a robust biological defense program. How does she reconcile the discrepancy? "I have mixed feelings. I think it's worth trying to develop detectors for the 'old' biological warfare agents like anthrax." But if genetically engineered organisms are used, "we are up against the wall."

Apart from the issue of defense, are Bailey's views about the futility of biological or chemical verification in the minority? "A tiny minority," she chuckles. The minority includes Frank Gaffney, director of the Center for Security Policy, a conservative think tank. A group of 50 contributors and outside grants support the center's annual output of 200 articles and papers. Gaffney:

If there had been no BWC [Biological Weapons Convention], it seems indisputable that the United States would have focussed far greater public, congressional and military attention on the Soviet BW program and the proliferation of this dangerous technology elsewhere around the world. . . .

In the presence of a treaty, however, anyone who was inclined to call attention to such developments was sharply challenged—not on the basis that the evidence was inadequate—but on the basis that they were enemies of arms control.[5]

For Rosenberg and Mahley a strengthened Biological Weapons Convention is hope. For Bailey and Gaffney it is anathema. How did we arrive at this juncture?

News watchers were kept busy in April 1972. The Vietnam War was bloodier than ever. Senator George McGovern was working toward the Democratic party's presidential nomination. American astronauts were en route to the moon. But while these events were amply covered in the media, another important story was virtually ignored.

When the *Convention on the Prohibition of the Development, Production and Stockpiling of Bacteriological (Biological) and Toxin Weapons and on Their Destruction* was opened for signature on April 10th, 79 nations signed immediately.[6] *The Wall Street Journal* made no mention of the occasion, nor did *Time* or *Newsweek*. Not until six days later did *The New York Times* note in a brief article that "last week a major step was taken" to outlaw biological weapons:

> More than 70 nations, including the United States, the Soviet Union and Britain, signed a treaty binding countries "not to develop, produce, stockpile or otherwise acquire or retain" biological agents except for peaceful purposes. And for the first time under a modern arms-control measure, the treaty requires states to destroy their stocks of such weapons.[7]

The *Times* piece was tucked inconspicuously in a section reviewing the week's news. Did not the first international agreement to eliminate an entire weapons system deserve more attention? Failure to highlight the story may have resulted from a sense of anticlimax. The terms of the new Biological Weapons Convention were revealed the previous October when a draft was submitted to the U.N. General Assembly. Moreover, the convention would not enter into force until 22 governments ratified, which happened in 1975.

But the treaty became a reality in 1972. If the moment was underappreciated at the time, it has since become a watershed. For admirers and critics, the treaty is a reference point in discussions about banning biological and chemical arms.

Banning Biologicals by Treaty

Perhaps the most important step toward establishing the 1972 convention took place two and a half years earlier. On November 25, 1969, President Richard Nixon announced several arms initiatives. Although the United States had not ratified the 1925 Geneva Protocol, which prohibited the use of chemical or biological weapons, he reaffirmed that we would never be the first to use lethal chemical weapons. He extended the abnegation to incapacitating chemicals as well.

Most dramatic of all, Nixon declared that existing stocks of biological weapons would be destroyed. The United States, he said, was renouncing "any form of deadly biological weapons that either kill or incapacitate."[8]

Three months later, the president added that toxins produced by microorganisms also would be eliminated. Only limited quantities of biological and toxin materials required for defensive research would be permitted. Moreover, he hoped other governments would follow the U.S. example.[9]

What prompted Nixon's action? According to David Beckler, chief of staff of the president's Science Advisory Committee in 1969, a recommendation by the scientists was instrumental. Committee members questioned whether biological weapons would be effective. Moreover, their use might cause disease to many noncombatants. Similar conclusions were drawn by other agencies, although the reviews are classified.[10]

In his renunciation statement, Nixon underscored humanitarian concerns, emphasizing that biological warfare might produce global epidemics. Some people believed his motives were less noble. One thought his intention was to prevent weaker countries from cutting into America's conventional and nuclear muscle.[11] Others saw him reacting to a growing domestic opposition to biological and chemical weapons.

News stories revealed that chemical warfare agents were leaking from storage containers. In 1968, a nerve-gas accident in Utah killed 6000 sheep. But even observers who thought Nixon's motivation more political than compassionate called his action "a brave gesture."[12]

Within two years, the U.S. initiative produced the intended effect. Even before Nixon's 1969 announcement, the United Kingdom and other Western countries had expressed support for an international treaty renouncing biological warfare. The Soviet Union resisted at first. It rejected the notion of a treaty that failed to deal with both chemical and biological arms. But in April 1971, the Soviets changed their position and presented a draft biological treaty to the Conference on Disarmament in Geneva. One year later, the Biological Weapons Convention was opened for signature. In the words of the convention, the "states parties" are

> determined, for the sake of all mankind, to exclude completely the possibility of bacteriological (biological) agents and toxins being used as weapons, [and are] convinced that such use would be repugnant to the conscience of mankind and that no effort should be spared to minimize this risk.

The document aims to fulfill its goals by encouraging consultation and information exchanges among states. Allegations of violations may be brought to the UN Security Council, but the council is empowered only to investigate and report its findings. Absent from the treaty are provisions to verify compliance. Nor are penalties provided for countries that cheat.

Assessments of the Biological Weapons Convention (BWC) variously stress its promise and disappointment. Two decades after the signing, University of Michigan science historian Susan

Wright called the convention "a major achievement in the history of disarmament." Conversely, Hugh Crone, a scientist with the Australian Department of Defense, saw it as a mistake now held in "almost universal poor regard." Both admirers and critics emphasize the absence of verification measures. To Wright, this absence is a weakness in need of correction, not an impairment of the treaty's magisterial uniqueness. To Crone, the absence has made the treaty disreputable.[13]

In an expansive review of the BWC through the mid-1980s, Nicholas Sims's conclusions are mixed. A lecturer at the London School of Economics and Political Science, he assesses whether the treaty has minimized the chances of a biological war. Until 1980, the answer was "wholeheartedly in the affirmative." Since then, however, the matter has become murky.[14]

ALLEGING SOVIET NONCOMPLIANCE

Beginning in 1980, the United States and a few other countries accused the Soviet Union of violating the BWC. Initially the claim was based on a 1979 outbreak of anthrax in Sverdlovsk, a city of 1.2 million. United States intelligence sources believed the epidemic had been caused by an accidental release of bacteria from a military facility. Hundreds of residents reportedly died of the disease in a period of weeks. The Soviets said the outbreak was caused by anthrax-contaminated meat and was unrelated to biological warfare.

Soviet authorities refused requests for an on-site investigation by outsiders. They said the incident was a matter of public health and not subject to the requirements of the Biological Weapons Convention. While regretting their inability to visit the area, some Western observers concluded that the Soviet contentions may have been accurate. Subsequent briefings to U.S. scientists by Soviet health officials convinced many that the Soviet version of the epidemic seemed "credible" and "plausible."[15]

In 1992, the year after the Soviet Union was dismantled, Russian President Boris Yeltsin acknowledged that the Soviet contentions were fabricated. The Sverdlovsk incident had indeed resulted from an accidental release of anthrax bacilli from a military research facility. Whether the research had been part of an illegal offensive program or a permissible defensive program remained unclear.[16]

Meanwhile, other allegations about Soviet biological activity soon surfaced. Late in 1981, U.S. officials accused the Soviets and their surrogates of using mycotoxins from the fungus *Fusarium* as weapons in Southeast Asia and Afghanistan. A State Department spokesman said that clouds of the toxins, described as yellow rain, were released from airplanes over defenseless people. Several scientists outside the government expressed skepticism. Tests in U.S. laboratories of blood, urine, and tissues of purported victims as well as samples of vegetation from the areas in question largely failed to reveal traces of toxins.[17]

Surmising that yellow rain might be nothing more than bee feces, Harvard biochemist Matthew Meselson led a team of scientists in Thailand to investigate. After examining material commonly dropped from Southeast Asian bees in flight, the scientists concluded that "the showers and spots closely resemble the showers and spots said to be caused by yellow rain."[18] By the end of the decade, the U.S. position had not changed. But few scientists outside the government were convinced that yellow rain was a toxin weapon.

These disagreements underscored the weakness of the Biological Weapons Convention. The Sverdlovsk and yellow rain issues, along with suspicions that the Soviets were developing genetically engineered weapons, caused some to question the value of the treaty. Without a means to verify, doubts about compliance would persist. The experiences prompted calls for strengthening the biological convention.

TRYING TO PATCH THE BWC

Review conferences have been held every five years since the convention went into effect. At the first conference in March 1980, delegates sought to encourage consultation among the states and to enhance confidence in the effectiveness of the convention. But division between several Western countries and the Soviet bloc was also evident.

The U.S. suspicions about the Sverdlovsk incident had surfaced just before the conference. The Swedish delegation led an effort to create a permanent committee to investigate suspected violations. The Soviet delegate rejected the idea: "There was no reason to worry about problems which did not exist." The Soviets opposed changes in the convention because it "was operating admirably."[19]

By the time of the second review conference in September 1986, the charges about Sverdlovsk and yellow rain had crystallized. Confidence in the BWC was shaken. Paradoxically, the Soviet Union now favored a verification regime. But the Americans had become skeptical, concerned in particular that genetic engineering could be clandestinely conducted. In effect, the Soviets had changed positions with the United States. The delegates agreed to defer the matter.[20]

The second BWC review conference concluded with proposals for more confidence-building measures. These proposals amounted to unenforceable efforts to make the convention more user friendly. They included calls for contacts among scientists doing work related to the convention and for exchanging data on laboratory activities.[21]

By the third review conference in September 1991, the international climate had radically changed. The coalition forces had defeated Iraq six months earlier, and the Soviet Union was on the verge of extinction. The number of states parties to the BWC had grown to 118.

Nevertheless, hopes that the treaty would be dramatically strengthened were unrealized. Obstruction came from two sources. Many developing countries resisted altering the treaty unless they could be assured of tangible benefits for themselves. They wanted guarantees that transfer of industrial technology would not be impeded.

The other obstruction came from the United States. Disagreeing with its Western allies and several other countries, the administration insisted that the BWC was not effectively verifiable. In the end, the Americans agreed to a feasibility study on the issue. The review conference established an ad hoc group of governmental experts, called VEREX, to examine the matter.

Arms control expert Nicholas Sims noted that on-site inspections were "still a long way off for this treaty regime however acceptable they may have become in others."[22] He may have exaggerated. The Chemical Weapons Convention (CWC) will likely influence the BWC in this regard. If verification procedures in the new CWC prove successful, they are likely to encourage the establishment of a verification regime for the BWC.

Banning Chemicals by Treaty

In his book *1984*, George Orwell introduced "newspeak." In 1984, U.S. chemical weapons policy was more like "confuse-speak." Vice President George Bush brought a proposal to the Conference on Disarmament that provided for on-site inspections anytime, anywhere. At the same time, the United States was expanding its chemical arsenal and substantively ignoring Iraq's use of chemicals against Iran.

The Conference on Disarmament (or its predecessor bodies) had been meeting in Geneva since the 1960s to devise a treaty to ban chemical weapons. After nearly two decades of plodding, the 40-nation body was in rough agreement about a framework. But the issue of verification remained the principal obstacle. In 1982, the Soviet Union proposed that nations could request an inspection of suspected chemical weapons facilities but that a challenged state could refuse. The state would be obligated only to "give appropriate and sufficiently convincing explanations."[23]

In the next year, the United States proposed that states accept "a stringent obligation" to permit inspections. The Soviets then said they could allow for international inspectors at facilities where stocks were destroyed.[24]

This was followed in April 1984 by a dramatic U.S. initiative. The 35-page draft convention that Bush presented to the disarmament conference contained, he said, an "unprecedented open invitation verification proposal."[25] Indeed it did.

In the words of the proposal, a state party to the CWC "shall have the right to request, at any time, . . . an *ad hoc* on-site inspection." If the request is deemed legitimate by a fact-finding panel (a small representative body of the parties to the convention), then the challenged state must "provide access within 24 hours of the Panel's request."[26]

The Soviets dismissed the proposal. Several Western states expressed misgivings about the U.S. move, contending that it was unrealistic and invited the Soviets to make capricious demands. Others deemed it an election year ploy.[27] The British believed the issue was tied to the Reagan administration's efforts to modernize the American arsenal with binary munitions. (Binary weapons contain two separate chemicals that combine to become lethal shortly before reaching a target.) According to the United Kingdom, the U.S. proposal was "in large part an attempt to smooth the binary programme through the Congress rather than to further the chemical weapons negotiations."[28]

Nevertheless, a summit meeting in November 1985 between President Ronald Reagan and General Secretary Mikhail Gorbachev resulted in a joint pledge to work to conclude a convention. From then on, the Soviets became increasingly conciliatory.

In April 1986, the Soviet delegate to the conference said his country would accept inspections of the destruction of production facilities. A year later, the Soviets seemed near to accepting the U.S. proposal. Foreign Minister Eduard Shevardnadze said his delegation to the Conference on Disarmament "will proceed from the need to make legally binding the principle of mandatory challenge inspections without the right of refusal."[29]

Valerie Adams, who worked on chemical disarmament in the British Ministry of Defense, believed the Soviet concessions were impelled by several reasons. The Soviets worried about chemical weapons proliferation and the cost of a chemical arms race with the United States. Most significant, however, was the new attitude in the Soviet Union.[30] Under Gorbachev, the Soviets were engaged in pragmatic cooperation with the West, which included efforts toward arms limitation.

By the end of 1987, the "rolling text" of the proposed chemical convention, as the draft was called, had grown to 100 pages. Provisions were being introduced that went beyond anything comparable in any other treaty. States would be required to declare their chemical inventories. Dozens of chemicals were listed in the treaty that had to be destroyed or monitored. An on-site inspection regime was being formulated. Sweden proposed a dramatic confidence-building measure—to ban *all* training related to chemical warfare, including defensive training, according to one observer.[31] The idea was rejected because some felt a temptation to violate the treaty might be greater if an enemy lacked chemical protection.

The free flow of suggestions seemed an expression of will to establish an effective agreement. The rolling text continued to expand. Listing too many details threatened to make the treaty unworkable, but the changing attitude of the Soviet Union continued to generate optimism. The distance the Soviets had come was demonstrated by Shevardnadze in January 1989. At a Paris conference on chemical weapons attended by some 150 states, he offered a memorable acknowledgment:

> Over the past two years, our position has evolved in a radical way from manufacturing chemical weapons to abandoning their production altogether, from hushing up data on the existing stockpiles to publishing such data, from seeking to protect chemical production and storage facilities from the eyes of others to recognizing the concept of comprehensive verification and inviting foreign observers to watch the elimination of chemical weapons. . . . And should anyone say to us that we waited too long before stopping the production of chemical weapons and imposing other prohibitions on them, we would say: yes, we did wait too long.[32]

The Paris conference ended with a pledge to redouble efforts to conclude a chemical weapons convention. But speakers at the Paris conference, as at the disarmament conference in Geneva, seemed blind to one reality. In both forums, scarcely a critical word was heard about Iraq's behavior. While laboring through the 1980s over clauses and commas, the 40-nation disarmament conference ignored Iraq's pummeling of Iran and its own citizens with chemical weapons. This silence carried into the Paris conference, which itself had been prompted in large measure by Iraq's actions.

Even more bizarre, delegates to the Paris conference sat passively as Iraq offered an initiative that would stop progress on a chemical ban. Foreign Minister Tariq Aziz insisted that a call for eliminating chemical weapons be linked to a ban on nuclear weapons. Ignoring his country's illegal use of chemicals, he sought to deflect attention to Israel's "direct aggression" in 1981 against a "peaceful Iraqi nuclear reactor."

Rather than reproach the Iraqis, the audience listened respectfully as Arab countries echoed Iraq's theme. In fact, the Arab League had endorsed an Egyptian proposal the previous year to eliminate weapons of mass destruction from the Middle East. Now Egypt's Foreign Minister Esmat Abdel-Meguid reiterated a part of that message: "Any progress on banning chemical weapons is tied to the conclusion of a parallel ban on nuclear arms."[33] The conference concluded with a plea for "general and complete disarmament," which satisfied the Arab states. As in other international forums, Iraq's use of chemical weapons seemed an invisible issue.[34]

The notion of linking nuclear to chemical disarmament was among several difficulties that emerged during this period. The proposed chemical convention called for eliminating all chemical weapons within ten years. In September 1989, President Bush announced that during the first eight years, the United States would destroy 98 percent of its stockpile. Destruction of the remaining 2 percent would be delayed until "all nations capable of building chemical weapons sign that total-ban treaty."[35]

The two-percent declaration drew harsh criticism. In the first place, the meaning of "capable" was unclear. Technically, every nation is capable of producing chemical weapons. Moreover, U.S. retention of chemical weapons in the name of security could give an excuse to others to do the same.[36]

Another cause of concern derived from a bilateral pact between the United States and the Soviet Union. The two nations agreed on June 1, 1990, to reduce their chemical stockpiles sharply. They proposed as well that eight years after a Chemical Weapons Convention entered into force a conference of states parties be convened. The conference would consider "proceeding to the total elimination of all remaining chemical weapons stocks over the subsequent two years."[37]

The long-sought cooperation between the two superpowers was, paradoxically, breeding resentment elsewhere. After attending a meeting of the Conference on Disarmament in August 1990, Congressman Martin Lancaster was pessimistic. "There is a feeling among some nations," he said, "that the bilateral agreement suggests a lesser importance for a multilateral Convention. Some even suggest that the US and the USSR are 'ganging up' on the rest of the participants to force their position on unwilling ambassadors."[38]

Lancaster also mentioned other areas of disquiet. The "greatest impediment" to the conclusion of a chemical convention, he said, arose from another reversal of American policy. The United States was now

> reneging on our agreement—on George Bush's own 1984 draft treaty—to accept challenge inspections "anytime, anywhere, without the right of refusal." Our current position states that challenge inspections should not be allowed near certain installations in the interest of security.[39]

Lancaster noted that the British had successfully conducted several mock challenge inspections. They showed that security interests could be protected, and he regretted that "at this point we find ourselves in the unfortunate position of having to polish the United States' tarnished image as a somewhat reluctant participant in the chemical weapons talks."[40]

Iraq's invasion of Kuwait in August 1990 and the ensuing Gulf War profoundly affected the international mood. As recounted in Chapter 5, Iraqi behavior convinced many that Iraq's chemical and biological capabilities must be eliminated.

Postwar efforts to conclude the Chemical Weapons Convention also intensified. Terms were hammered out for many problematic issues including disposition of existing weapons, conversion of arms production facilities to industrial use, export controls, organization, and membership of the oversight bodies.

From beginning to end, verification was the most contentious issue. A British summary of the negotiations recounts that the verification regime was not concluded "until the final phase of the negotiations in the summer of 1992."[41]

In September 1992, the Conference on Disarmament adopted the Chemical Weapons Convention. In November, the UN General Assembly passed a resolution in support of the convention and requested the UN Secretary General to open it for signature.

Two months later, on January 13, 1993, at a gathering in Paris, delegates from 131 countries signed the *Convention on the Prohibition of the Development, Production, Stockpiling and Use of Chemical Weapons and on Their Destruction.* The formal title is almost the same as that for the Biological Weapons Convention. But the BWC is five pages long, the CWC 186.[42]

The Chemical Weapons Convention enters into force six months after 65 nations have deposited their instruments of ratification with the UN Secretary General. By early 1995, 159 nations had signed and 29 had deposited. The United States had not yet ratified but was expected to do so before the end of the year. Support had previously been voiced by spokesmen from every sector that could be directly affected—the Pentagon, the intelligence community, the Chemical Manufacturers Association, and arms control experts. The convention had been signed by the Bush administration and was supported by the Clinton administration. Seventy-five senators had previously signed a resolution in support of the convention. Needing a two-thirds majority in the Senate, ratification seemed assured.

By the end of 1995, early U.S. ratification no longer appeared certain. Senator Jesse Helms, chairman of the Foreign Relations Committee, announced opposition. For the Senate to consider the treaty, it first had to pass through his committee. This he was reluctant to permit because he doubted that the treaty could be effectively verified.[43] Despite his opposition, however, in April 1966 a resolution was voted out of the committee in favor of ratification. But a ratification vote was postponed indefinitely after majority leader Robert Dole left the Senate and criticized the treaty during his presidential campaign.

Verification

The treaties banning biological and chemical weapons clearly have affected each other. This is especially true in the area of verification. The absence of verification provisions in the 1972 Biological Weapons Convention proved to be a major weakness. Alleged violations by the Soviet Union and others in the 1980s could not be satisfactorily addressed. This reality influenced deliberations at the Conference on Disarmament on the chemical ban.

In large measure, the verification protocol in the Chemical Weapons Convention is attributable to the biological experience. The result is an elaborate system of detailed procedures. The chemical protocol in turn stimulated interest in developing a verification regime in the Biological Weapons Convention. Before reviewing the strides taken in the biological area, we examine the provisions for chemical verification.

THE CWC FRAMEWORK

Article VIII of the Chemical Weapons Convention describes three organs to fulfill its goals. The Conference of the States Parties, which is to meet at least once a year, provides overall policy direction to the treaty. Day-to-day administration is conducted by a smaller Executive Council of 41 states parties. It supervises the Technical Secretariat, which is responsible for the treaty's most sensitive function—carrying out "the verification measures provided for in this Convention." The measures include both routine and challenge inspections.

Requests for a challenge inspection are submitted to the Executive Council and the director-general of the Technical Secretariat. The director-general in turn must inform the challenged state at least 12 hours before an inspection team arrives. The procedure can be stopped if three-quarters of the Executive Committee deem the request "frivolous, abusive or clearly beyond the scope of this Convention" (Article IX).

As a preface to describing verification procedures, the CWC lists three categories of toxic chemicals and their precursors. The listings typify the convention's effort to minimize ambiguities. Schedule 1 includes agents that pose "high risk to the object and purpose" of the convention. Schedule 2 lists chemicals with a "significant risk" to the convention, and Schedule 3 comprises agents that "otherwise" pose a risk to "the object and purpose" of the convention. Examples in the first category are nerve agents such as sarin, mustard agents, saxitoxin (a toxin derived from shellfish), and ricin (a toxin from castor beans). Items in Schedule 2 include BZ (a hallucinogen), arsenic, and thiodiglycol. Schedule 3 includes phosgene and hydrogen cyanide (both used as chemical weapons in World War I) and a variety of precursors. Altogether, 43 items are named. Some are specific agents or precursors, and others are groups of structurally related chemicals (Annex on Chemicals, Schedules of Chemicals).

The designated materials form the basis on which nations must declare their stocks and production facilities. Many of the items are used for industrial or scientific activities. In declaring the quantities and locations of the materials, a state offers assurance that Schedule 1 chemicals "are applied to research, medical, pharmaceutical or protective purposes." Moreover, the aggregate amount of these chemicals at a given time may not exceed one metric ton (Verification Annex, Part VI).

For Schedules 2 and 3 chemicals, there are no ceilings on quantities. But states must declare annually the locations, purposes, and amounts of chemicals produced, processed, or consumed above certain thresholds (Verification Annex, Parts VII and VIII).

The significance of the verification issue is revealed by the amount of space given to it. The Verification Annex comprises 100 of the convention's 186 pages. The annex covers the designation of inspectors, their privileges and immunities, rules concerning the conduct of inspections, protocols for the destruction of old weapons, and the destruction or conversion of production facilities. The most sensitive section of the annex is part X, which deals with challenge inspections.

Part X is filled with details, like the fine print in an insurance policy. But the pioneering message brings the prose to life—a framework for intrusive inspections across national borders. The verification protocols constitute a daring reach. They amount to a codified breach of national sovereignty.

Any state party may request an inspection of another state party. The request is expected to specify the perimeter of the site to be inspected. After notifying the challenged state, the head of the Technical Secretariat will dispatch an inspection team. Within 36 hours of arrival in the challenged state, the team must be transported to the site perimeter.

The inspectors then secure the site by verifying that "there is no exit activity." If the inspection team is denied full access, the inspected party must demonstrate that its actions are unrelated "to the possible non-compliance concerns raised in the inspection request."

The inspection must begin no later than 108 hours after the team arrives in the country. The convention recognizes a country's right to protect national security, proprietary information, or other legitimately concealed information. The obligation is worded cautiously:

> If the inspected State Party provides less than full access to
> places, activities, or information, it shall be under the
> obligation to make every reasonable effort to provide
> alternative means to clarify the possible non-compliance
> concern that generated the challenge inspection.

Negotiating and implementing the means of clarification are described as "managed access." To prevent disclosure of confidential information, the convention allows an inspected state a variety of actions. They include removing sensitive papers from office spaces, shrouding sensitive displays and equipment, logging off computer systems, and restricting sample analysis to the presence or absence of chemicals listed in the schedules. But to convince inspectors that the measures are not intended to hide noncompliance, the convention suggests, for example, "partial removal of a shroud" that could satisfy such concerns (Verification Annex, Part X).

What happens if a state is found in violation of the convention? On recommendation of the Executive Council, the Conference of States Parties may seek "collective measures" conforming with international law. In cases of "particular gravity," the conference of states shall bring the matter to the United Nations, which may decide to impose sanctions (Article XIV).

TOWARD BIOLOGICAL VERIFICATION

Verification is more difficult for biological materials than for chemicals. Scores of pathogens that could be used as weapons are natural substances. Moreover, they are commonly used in industry, hospitals, and academic and research laboratories. Providing assurance that they are intended for permitted purposes seems daunting. But by the end of the 1980s, more observers began to consider the possibilities.

In 1990, Marie Chevrier, an arms control analyst, addressed the matter in an article intriguingly titled "Verifying the Unverifiable: Lessons from the Biological Weapons Convention." Acceptable verification need not be leakproof, she wrote. Describing her criterion as "adequate verification," a regime would require only that cheating be detectable at a level that might threaten U.S. security. This could be achieved by consultation between states, exchange of information, and voluntary on-site inspections. Chevrier supported building confidence with such incremental steps rather than amending the treaty.[44]

Meanwhile, the third review conference of the Biological Weapons Convention, held in 1991, proposed that the verification issue be examined. As mentioned earlier, a group of governmental experts was established to evaluate potential measures. Their mandate was limited to examining scientific and technical questions. VEREX, as the group was called, produced a report in September 1993.[45]

The primary VEREX contribution was to identify 21 potential verification measures. They range from the distant to the intrusive. Among the more general and less effective means are surveillance of publications and legislation and sharing of information. Surveillance by satellite, aircraft, or other off-site devices might have some value but was also seen as unlikely to resolve ambiguities.

The most effective means of verification are more intrusive. They include on-site visual inspections, sampling, identification of key equipment, interviewing, and continuous monitoring by instruments and personnel.[46]

Because of its limited mandate, VEREX did not recommend which techniques to include in a verification regime. But VEREX members clearly believed verification possible:

> [T]he Group considered, from the scientific and technical standpoint, that some of the verification measures would

contribute to strengthening the effectiveness and improve the implementation of the Convention, also recognizing that appropriate and effective verification could reinforce the Convention.[47]

A majority of states decided to go further. A special conference was convened in September 1994 to consider the VEREX report. Some states were enthusiastic about verification, and others were not. On behalf of the European Union, Germany declared that a "verification protocol should now be concluded as expeditiously as possible."[48] Similarly, the Australian delegate believed that the VEREX report mandated the drafting of a verification protocol and that "falling short of this goal would represent a failure."[49]

But some developing countries expressed reservations. The Indonesian delegate spoke for several who worried that verification might impede biotechnology development and infringe on their national sovereignty.[50]

The United States also was hesitant. In a statement to the conference, Donald Mahley, head of the delegation, spoke of strengthening the Biological Weapons Convention. But not in the context of verification. Rather the United States emphasized the need for "transparency" by declarations of facilities, stocks, and activities. Scrupulously avoiding the word "verification," Mahley said the process should consider only on-site or off-site "measures."[51]

Despite these differences, the 1994 Special Conference agreed to establish an ad hoc group that would propose "a system to verify effective compliance." Moreover, the proposal "shall be submitted, if possible [for endorsement] by the Fourth Review Conference in 1996."[52] To this end, the ad hoc group met several times in anticipation of the review conference, scheduled for December 1996.

TAKING VERIFICATION SERIOUSLY

The notion of verification is daunting in view of the number and variety of potential agents. At a nongovernmental forum in conjunction with the 1994 Special Conference, a paper by a Brazilian official identified 148 viruses, bacteria, rickettsiae, fungi, and toxins that might be used as weapons. All are naturally occurring. Novel genetically engineered agents were not even considered.[53] But the paper stood as a signal of increasing momentum in the effort to achieve a verification protocol.

Another paper, by the Federation of American Scientists (FAS), estimated costs of a verification regime. The figures were based on "related precedents" and assumptions. Among thousands of "relevant facilities," only about a hundred might require inspection in a year. Each inspection would require five inspectors and average two and a half days. The paper delineated costs for personnel, training, transportation, equipment, laboratories, and other items. The annual total would be $14.5 million.[54] Assuming the FAS projections to be realistic, the cost is surprisingly modest. In any case, they provide a valuable base line for discussion.

In the end, whether for biological or other arms agreements, the underlying question about verification is the same: Are the benefits likely to exceed the risks? In 1989, Robert Mikulak, a senior scientist with the U.S. Arms Control and Disarmament Agency, wondered what level of technical assurance is acceptable: "Ninety percent? Fifty percent? Twenty percent? Where do you draw the line?"

Mikulak's inquiry was rhetorical because the answer is not quantifiable. Nevertheless, "although the monitoring question cannot be solved in the sense of having absolute verification, one can get a high enough level of verification to detect clandestine facilities through a variety of means—not only through inspection but through intelligence means."[55]

Five years later, Michael Moodie, president of the Chemical and Biological Arms Control Institute, noted that verification "is not a mechanistic, cut and dried process that produces unambiguous evidence of noncompliance." A single inspection may not produce conclusive findings about noncompliance. Rather, judgments derive from "a mosaic of evidence created over time from a range of activities including inspections, interviews, evaluations, and non-treaty related inputs."[56] Others have observed that raising verification requirements to "unreasonably high levels" can jeopardize any arms control process.[57]

Vil Mirzayanov, a scientist who worked in the Soviet chemical weapons program, believed illegal weapons development might still be going on in Russian laboratories. "If the CWC's procedures are not instituted, the Russian chemical weapons complex will remain accountable only to the same clique of leaders, who have thus far not proven their trustworthiness."[58] For Mirzayanov and other experts, the verification protocol addressed their deepest concerns.

Similarly for biological weapons. Despite President Boris Yeltsin's 1992 admission that the Soviets had run an offensive program, Western governments suspect that illegal biological work continues in Russia.[59] Moreover, recent recruitment of scientists, including biologists, from the former Soviet Union by Libya and Iran has concerned U.S. officials.[60] Are the biologists being hired to develop weapons?

A strengthened Biological Weapons Convention would help answer these questions. Although it is difficult to identify illegal biological work, a verification regime can go some distance. Knowledge that a short-notice inspection can be demanded will itself discourage illegal weapons activities. The agreement also can be expected to offer trade and educational advantages in biotechnology to member states. Thus a state that fails to become party to the agreement not only risks being tainted in the general community, but deprives itself of economic benefits as well.

Conclusion

The wisdom of any position can be confirmed only in retrospect. The Chemical Weapons Convention will have to have been in force; a verification protocol for the Biological Weapons Convention will have to have been established. Even then, results may be ambiguous.

Inspections of a state may fail to reveal violations. Yet later discovery might show that the state was hiding biological or chemical weapons. Supporters of the conventions will insist that the treaties were never advertised as foolproof. They will say that the offending action was exceptional, that other states remain in compliance, that ultimately noncompliance was exposed.

Opponents of the conventions will argue that the incident validates their skepticism. They will stress the initial failure to detect the violations. Other states may be harboring illegal weapons as well, they will say, and we are fooling ourselves to believe otherwise.

Both sides will pronounce their earlier positions vindicated.

Understanding that ambiguity is likely, however, should not preclude seeking the wisest policy. Apart from their rejection of the potential moral influence of the conventions, opponents fail on more narrow grounds. Frank Gaffney inadvertently provides several examples. If not for the Biological Weapons Convention, he said, "it seems indisputable" that the United States would have addressed biological weapons proliferation. His contention is indeed disputable. How would the United States have stopped the growing number of programs? He does not say.

The most compelling explanation for the spread of biological and chemical weapons has nothing to do with the convention. It arose from the disregard by the United States and others during the 1980s of Iraqi behavior. The administration that Gaffney was serving was letting Iraq get away with chemical murder.

Gaffney and Kathleen Bailey think verification a waste of money. No one can "prove" that they are incorrect, although the chances that they are seem well worth taking. An expansive civilian defense program, which they endorse, is much more likely to be a waste of money. A verification regime bears uncertainty, but far less than a civilian defense program.[61]

The greatest value of strong chemical and biological weapons conventions is more than their structural provisions. The treaties are affirmations by the world community that these weapons are illegitimate. That nations are, in the preambles of both conventions, "determined for the sake of all mankind, to exclude completely the possibility" of their use.

Strengthening international agreements provides the most important response to states that contemplate chemical and biological weapons programs. The agreements will serve to renew contracts that have become stale. They will reinvigorate the sense of moral repugnance about these weapons that was weakened in the 1980s. And they establish the legal justification for action against parties that violate the norm.

Without strong treaties, proliferation is likely to continue. With strong treaties, the chances of transgressions are reduced. In arguing for ratification of the chemical weapons treaty in 1994, Amy Smithson, of the Henry L. Stimson Center, asked the preeminent question: "If not the convention what?"[62]

BIOLOGICAL WAR, BIOLOGICAL PEACE

Peace through arms. Peace through disarmament. Through strong defense. No defense. Condemnation. Economic sanctions. Prayer. All the notions have been embraced and scorned. Dilemmas about competing advocacies in the name of peace are nowhere better demonstrated than in the area of biological warfare.

By the early 1980s, the integrity of the Biological Weapons Convention was under challenge from contrary directions. The U.S. administration alleged that the Soviets were operating an offensive program in violation of the treaty. At the same time, critics suggested a potential for treaty violations in the activities of the United States itself.

These attitudes ripened into opposing schools of thought among observers. One, grounded in a military solution, stressed the need for stronger defenses against biological agents. The other saw risks in the U.S. Army's biological defense program. It urged civilian control of defense research and a verification system to assure treaty compliance.

Few people are absolutists. Many who emphasize defense also support verification. Nor do enthusiasts of civilian control believe that military forces should be unprotected. But some advocates lean heavily toward one or the other approach. Although structured on the dichotomy, the discussion begins with a broader, centrist framework offered by Graham Pearson.

Pearson's Web

In 1992, Graham Pearson, then head of Britain's Chemical and Biological Defense Establishment, began to use with some frequency the expression "web of deterrence" in discussions about preventing biological or chemical warfare. His web initially consisted of three strands: arms control with verification, export controls, and protection.[1] Later, he added a fourth: responses to acquisition or use.[2] Pearson was so fond of his euphonic phrase that he used it in at least seven articles and papers. On biological weapons, for example:

> National and international security requires that all nations do all that they can to discourage states from contemplating the acquisition of biological weapons. The way to achieve this is through a web of deterrence comprising effective protective measures that will reduce the utility to an aggressor of biological warfare, a strengthened Biological and Toxin Weapons Convention with an effective intrusive verification regime, broad export monitoring and controls making acquisition of biological warfare materials and equipment more difficult, and determined national and international responses to noncompliance.[3]

Pearson's four strands have received mixed support. The United States likes the strand on export controls though not the one on verification. Conversely, the Federation of American Scientists opposes export controls but wants verification. (The organization questions whether export controls would be effective and believes that opposition by some countries to export controls has impeded their endorsing a verification protocol.[4])

The third strand, response, includes "the possibility of an armed response—if a state is found to be acquiring biological weapons or has gone so far as to use them."[5] The idea is demonstrably valuable. The threat of retaliation evidently deterred Iraq from using biological and chemical arms in the 1991 Gulf War. But the will of the international community to impose sanctions or use force has historically been inconsistent. The strand will be effective only if punishment for treaty violations seems likely.

The remaining strand, protection, touches on two long-standing controversies about biological defense: the offense-defense dilemma and the limitations of protecting a large population.

The Offense-Defense Dilemma

In the realm of disarmament, 1969 was a watershed year. Richard Nixon's renunciation of biological weapons anticipated the 1972 Biological Weapons Convention, the first treaty to ban an entire category of weapons of mass destruction. But by preserving a defensive program, the United States has been forced into verbal acrobatics ever since.

Until Nixon's announcement, offensive and defensive research were officially considered almost the same. From a 1949 Pentagon report: "Information obtained from research on the defensive aspect of BW is, in the greater part, applicable to offensive problems as well."[6] As recently as 1968 an army publication indicated that "research and development in the offensive aspects of BW proceeded hand in hand with defensive developments for, in truth, the two are almost inseparable."[7]

After 1969, the official position turned upside down. Now defensive work was proclaimed distinctive and separable. "The only similarities between offensive and defensive research are that common laboratory techniques are used in each at the outset," claimed army spokesmen, "but even at the outset, the experimental hypotheses are diametrically opposed."[8] The idea that U.S. defense research in biotechnology might be used for offensive purposes was dismissed as "patently absurd."[9]

Which is correct, the army's pre-1969 wisdom or its post-1969 reversal? To Massachusetts Institute of Technology scientists Jonathan King and Harlee Strauss, the answer is indisputable: "It should be noted that the blurring of defensive and offensive programmes is not limited to research, but extends to development, testing, production and training."[10]

Strauss and King examine a cornerstone of defensive research —the development of vaccines—which requires production of biological agents. The agents are necessary not only in the formulation of vaccines but in testing their effectiveness. An agent must be grown and induced to infect animal and human subjects. The resulting information would be of no less value to an offensive program than to a defensive one.[11]

The 1991 Gulf War brought the dilemma to life. Before and during the war, the United States cultivated large amounts of botulinum toxin, which was necessary to make a vaccine. But until the toxin was inactivated, no one could be sure whether it was being produced for offensive or defensive purposes.[12]

Several additional difficulties relate to vaccines as a defense in biological warfare. A full immune response may not develop until months after a vaccination is administered. Even then, high exposure to a pathogen may overcome presumed protection. The most imposing problems are the diversity of biological warfare agents and the possibility that an enemy has altered their genetic structures.[13]

The army cannot reasonably dismiss these observations, because they are true. Rather, spokesmen from army laboratories at Fort Detrick, Maryland, shift the focus. They emphasize the army's obligation to protect troops against biological weapons "by every medical means available." To this end, the army has already developed vaccines against several potential agents.[14]

A West Point professor, Major Michael Frisina, expands on the theme. In an article titled "The Offensive-Defensive Distinction in Military Biological Research," he holds that a vaccine research program run by the military "is a pragmatic and moral necessity."[15] He differentiates between vaccine research and weapons production:

> To obtain a vaccine, only small amounts of an attenuated virus are necessary for testing efficacy, conducting preclinical trials, and safety testing, production and distribution. Weapons production follows an entirely different pathway. The virus is made more virulent, stabilized, produced in large quantities, and finally a weapons and delivery system must be produced and field tested.[16]

Frisina's distinction between offensive and defensive programs is based on two criteria. One is the quantity, virulence, and stability of bioweapons being produced; the other whether delivery systems are being tested. He supports the production of pathogens in small amounts to develop vaccines and to test against detector systems, protective gear, and shelters.

"Small" quantities may have significance when describing other weapons systems, but hardly with biologicals. Unlike other weapons, bacteria and viruses can reproduce themselves. Under the right conditions, a small number of stable, highly pathogenic organisms can become large in a few days. In this regard, the distinction between defensive and offensive work can be so transient as to be negligible.

The presumption that an offensive capability requires the testing of delivery systems also is questionable. This might be true if the systems were limited to munitions and shells. But the presumption ignores the history of army testing that demonstrated the effectiveness of other methods.

From 1949 through 1969, the army simulated hundreds of biological warfare attacks across the country. Bacteria and chemicals were sprayed from boats, from automobiles, from suitcases, from light bulbs tossed on subway tracks. Delivery equipment, whether fans or glass containers, could be purchased at a hardware store. Indeed, in 1996, an army spokesman emphasized that "complex systems" of delivery were unnecessary. "Something as simple as a pesticide sprayer can be used to spray biological agents."[17]

None of this means the army is now engaged in illegal activity. But the inclination by officials to minimize the overlap between offensive and defensive work provokes unease. Suspicions were fueled by a 1993 report that army-sponsored researchers were developing an antibiotic-resistant strain of anthrax and a vaccine-resistant form of botulism toxin.[18]

On one matter everyone agrees. Because offensive and defensive programs share the same components, a key word is "intent." The complexity of the issue was highlighted in a 1992 book on the ethics and politics of biological defense research:

> The general criterion for distinguishing between offensive and defensive research is *intent* [original emphasis], which at best is a problematic issue. We often assume that if defense research is carried out openly, has adequate

oversight, and that the results are published, its intent is peaceful. But how does a scientist know that these safeguards will be adhered to when he or she undertakes Department of Defense-sponsored research? More importantly, do these safeguards really ensure that the research will be used only for peaceful purposes?[19]

Contributors to the book reflected a range of views: that exclusive defense research is possible, that overlap exists only in the initial stages of research, or that "intent is difficult, if not impossible, to verify."[20]

The most foreboding comment came from historian Sheldon Harris, an expert on Japan's biological warfare experiments in the 1930s and 1940s. In the past, when states said they were doing purely defensive research, the work "led inevitably to offensive biological warfare research."[21] In thinking beyond the experience of the Japanese or Germans during that period, Harris's generalization of "inevitability" seems overdrawn. But his concern deserves respect and is a reminder of the need for unrelenting caution about *any* research in this area.

Is Defense Possible?

Is protection possible against a biological attack? Yes, in certain situations. The agent must be accurately identified. It must be susceptible to a vaccine that was administered in advance of exposure. If antibiotics or other medications can offer effective treatment, they must be given soon after exposure. If the agent's potency is short lived, respirator masks may offer protection if donned before exposure. (A chemical attack requires a more cumbersome mask and outerwear. But chemically poisoned environments are also more amenable to early detection and decontamination.[22])

Under prescribed conditions, therefore, protective efforts may be effective. To the extent such measures can help soldiers in battle, they should be available. But in a variety of scenarios defense seems improbable, especially for a large population.

VACCINATION

Critics have written that the biological defense program is "highly ambiguous, provocative, and strongly suggestive of offensive goals."[23] Moreover, the boundless variations of possible biological warfare agents mean that immunizing a population against an attack may amount to self-deception.

This has prompted calls to eliminate the army's vaccination research. The risks associated with the effort are seen as not worth any value that might come of it. Victor Sidel, former president of the American Public Health Association, believes government-sponsored medical research for "prophylactic, protective, and other peaceful purposes" should be under the auspices of civilian agencies such as the National Institutes of Health and the Centers for Disease Control.[24]

The Boston-based Council for Responsible Genetics (CRG) agrees. Indeed, in the late 1980s, the CRG distributed a petition against secret biological research. Warning that genetic manipulation under auspices of the army's program could lead to "the eventual production of new biological weapons," the petition drew 4000 signatures. By 1991, more than 1500 scientists had signed another CRG petition that asked biologists and chemists to pledge "not to engage knowingly in research and teaching that will further the development of chemical and biological warfare agents."

When protection of a target population is being considered, inoculations can be problematic even when the nature of an organism is known in advance. The threatened swine flu epidemic in the mid-1970s is instructive. Although the government urged citizens to seek vaccinations, few complied. (In the end, the epidemic never materialized, and several people who received vaccinations suffered serious side effects.) The public is no more likely to comply with a vaccine program against suspected biological weapons than against swine flu.

The difficulty in developing vaccines for some agents is nowhere better demonstrated than in the example of the AIDS virus. Since the early 1980s, investigators throughout the world have been seeking a cure and a vaccine for AIDS. Between 1985 and 1995, the United States spent more than $10 billion on AIDS research, nearly 10 percent of which was spent on searching for a vaccine.[25] The search has dwarfed efforts to find a vaccine against any biological warfare agent. Yet no prospective AIDS vaccine has thus far proved effective.

Apart from vaccines, protective clothing or shelters can be helpful only for limited periods. If an organism becomes established in an ecological niche, conditions can remain as dangerous and intractable as a radiation-filled environment after a nuclear attack. The usual concept of defense in such a situation becomes meaningless.

But vaccines and protective gear are not the only challenges in biological defense. Identifying an organism quickly in a battlefield situation also is problematic. Even determining whether a biological attack has been launched can be uncertain. In consequence, the Pentagon has begun to pay more attention to detection.

DETECTION

In May 1994, Deputy Secretary of Defense John Deutch produced an interagency report on counterproliferation activities concerning weapons of mass destruction. Biological agent detectors in particular, he wrote, were "not being pursued adequately." To the annual $110 million budgeted for biological and chemical weapons detection development, the report recommended adding $75 million.[26] Already underway were Pentagon-sponsored programs with such exotic names as ion-trap mass spectrometry and laser-induced breakdown spectroscopy.[27]

A year later, the army's coordinator of biological defense projects presented an overview of progress. At 55, Brigadier General Walter Busbee stood trim and confident. The office he managed, the Joint Program Office for BioDefense, was created in 1992, after the biological scare during Desert Storm. "What we want is the technology to interrogate and identify particles through mass spectrometry, some means of ionization or fragmentation or breakdown." Busbee was referring to a technique that could identify a range of biological agents. By developing a "library" from the fragments and their ionization products, "we may be able to come up with the 'silver bullet'."[28]

Busbee's silver bullet would be a generic detector of biological agents. The goal, he granted, is a long way off. Meanwhile, the military is also advancing a more limited approach that identifies specific agents through antibody-antigen combinations.

A cloud of anthrax bacteria is invisible, he pointed out, and people would not realize they were inhaling the organisms. Symptoms do not appear until 24 to 72 hours after exposure. If untreated, 99 percent of the people exposed would die. But if administered during the first 24 hours, antibiotic treatment is nearly "100 percent effective for people inoculated before exposure." (Without prior immunization, even early antibiotic treatment may be unhelpful.) No wonder military officials would like a detection system.

How close are they to one? "Currently we have nine prototypes of a BIDS system," Busbee said. BIDS, which stands for Biological Integrated Detection System, is a system on wheels intended to warn a soldier in the field. Interim and long-range warning systems also are under development, Busbee continued.

A BIDS apparatus first tries to establish the presence of biological materials through generic methods. They include the use of instruments to determine particle size, the presence of nucleic acids, and energy processes that indicate biological activity. A suspected sample is then exposed to antibodies of a particular biological agent, such as the anthrax bacillus. A reaction of the antibody would signify that the anthrax agent is present, a process that takes about 30 minutes. According to Busbee, BIDS can now identify four agents through antibody-antigen reactions: *Bacillus anthracis,* botulinum toxin, *Yersinia pestis,* and *Staphylococcus* enterotoxin B. He expects that by 1997 four more agents will be detectable: ricin and the organisms that cause cholera, tularemia, and brucellosis.

A government consultant who requested anonymity is less optimistic. He saw a BIDS apparatus and "was amazed by how crude and cumbersome it looks. It's not clear to me that it is effective in identifying the four agents it's supposed to. And we've spent how-many-millions on this already?"

Laboratory investigations to identify additional agents through antibody-antigen reactions are underway, Busbee said. But scores of organisms and toxins are viewed as potential biological warfare agents. Whether the full range, or even most, will be detectable by BIDS remains uncertain. Busbee acknowledged that the system "will be limited always to the current capability of antigen-antibody combinations in order to get identification of the items. That is troublesome." Indeed, no BIDS device is yet considered suitable for field use. Busbee hoped that small reliable systems would be available "within a decade or so."

BIDS and other detection systems are being tested at Dugway Proving Ground in Utah. Busbee mentioned that a helicopter-mounted laser can detect particles in a cloud released by an air sprayer. What kind of materials are being sprayed during the tests? Aware of citizen unrest about Dugway, Busbee is sensitive to such concerns (discussed in Chapter 4): "We're licensed and have approval from the state of Utah and the Environmental Protection Agency to use BG." The reference is to *Bacillus subtilis*. The bacterium, described by the army as a safe simulant, has long been used in outdoor testing. "In the past we have also had some tests with *Bacillus thuringiensis*, BT." A naturally occurring microorganism, BT is also used as a pesticide.

Busbee emphasized that biological warfare agents are not being sprayed outdoors. "We take the components of the detection systems and challenge them in a closed sealed chamber with BG or BT, or the real mechanism—anthrax, botulinum toxin." The instruments are then tested outdoors, using only simulants.

Do people wear masks or other protective gear when testing with simulants? "Yes, they use respirators."

Is there general notification when a test is being conducted so that others are aware? "The test range is fully closed off, instrumented with appropriate measures for not having people coming and going." The army's long-standing assumption evidently continues to be that released materials will remain confined to a chosen location. As demonstrated in Chapter 2, this assumption is not always warranted.

In September 1995, Dugway hosted the first annual "sniff-off," as Busbee described it, to test a variety of detection systems. He said that the army plans to hold sniff-offs every year to assess which systems are worthy of further development. The testing includes the release of huge numbers of bacteria. How comfortable the citizens of Utah will be about these post–Labor Day sniff-offs remains to be seen.

Busbee then commented on genetic engineering. "There's a potential for someone very smart with a lot of technology to take an organism and modify it slightly," he said. "So, in fact, our current antibody-antigen based detection modus can be tricked, or at least masked, so that you won't get a clear unequivocal signal." That will always be a problem, Busbee conceded, "until you get a generic detector which does not rely on that antibody technology."

Nevertheless: "If the question is should we be expending resources before we solve that technical challenge, I say yes." Countries with biological weapons programs appear to be using predictable agents, he continued, and for that reason alone protective efforts make sense.

One of the generic approaches he has in mind relies on the use of mass spectrometry. Harvey Ko, an electrophysicist, directs a Pentagon-contracted detection project at Johns Hopkins University's Division of Applied Physics. Ko thinks a mass spectrometry system will eventually be able to identify a broad range of organisms. But could not organisms be altered to avoid detection? "That's true," he answers, "but it doesn't impede our program." The nature of warfare means that the aggressor always has the first advantage, just as a naturally occurring pathogen does. "You try to stay ahead of your aggressor, just as we fight disease and try to stay ahead of the next flu virus."[29]

Optimists about the future of such detection instruments are matched by knowledgeable pessimists. Will Happer, a Princeton physicist who headed an advisory panel to the Pentagon, thinks these detection systems are unlikely ever to work in real battle conditions.[30]

What, then, is proper defense policy? Improbable as comprehensive defense against a biological attack is, under some conditions protection is possible. If the concept of defense is appropriate, then doing away entirely with defensive research is unrealistic. As long as the risk of biological attack exists, improving gas masks or developing vaccines is legitimate. But to ignore their limitations, or for that matter the potential for excesses in the U.S. defense program, is to ignore reality.

The most effective protection against biological warfare is, and will be, prevention. Apart from strengthening the Biological Weapons Convention, two other proposals with this aim have surfaced in recent years.

Surveillance, Vaccines, and Peace

Kühlungsborn is a lazy resort town on the Baltic coast of the former East Germany. In September 1990, two weeks before German reunification, some 30 scholars and diplomats from throughout the world met there for a colloquium on preventing a biological arms race.[31] The conference became the springboard for two ideas that subsequently attracted broad attention. One, from University of California microbiologist Mark Wheelis, concerned epidemiological surveillance. The other, from Erhard Geissler, a geneticist who organized the conference, became called "Vaccines for Peace."

GLOBAL EPIDEMIOLOGICAL SURVEILLANCE

Why not, Wheelis asked, create a system that could rapidly detect disease outbreaks throughout the world? Global epidemiological surveillance could help establish whether an outbreak was caused by a hostile source. The system would thus enhance compliance with the Biological Weapons Convention not only by detection, but by deterrence.[32]

Wheelis called for a four-pronged approach: a system of reporting to detect epidemics; mechanisms for prompt field and laboratory testing; analytic capability to deduce origin; records to maintain a base line and to understand worldwide patterns of disease.

How to determine that an outbreak had hostile rather than natural origins? Wheelis said "a number of suspicious circumstances" might suggest clandestine biological warfare. The causative agent may have been known previously only from culture collections or its genetic patterns might suggest manipulation. Suspicions could arise if a usual means of transmission is absent (such as insects or mosquitoes) or if the disease is caused by unexpected routes (e.g., inhalation of organisms not commonly airborne). Unusually high fatality rates or odd distribution patterns also would be suspect.[33]

Wheelis suggested that the system could be administered by existing organizations. Such United Nations–associated agencies as the World Health Organization (WHO) and the Food and Agricultural Organization (FAO) were already suited to monitor human, animal, and crop diseases.

Commentaries on Wheelis's ideas largely supported his concerns but not his emphasis. Their gist was that an epidemiological surveillance program should be established to understand disease patterns in general. At the same time, strengthening the biological weapons treaty would be a desirable side effect.[34]

Wheelis took the advice. In 1994, he wrote that, although his idea "grew out of weapons control considerations, it is clear that the principal benefits of such a system would be in the arena of public health more generally." They would come from expanded clinical facilities needed for diagnosis and earlier detection of emerging diseases.[35]

By 1995, Wheelis had joined his effort to a project sponsored by the Federation of American Scientists called ProMED (Program to Monitor Emerging Diseases). Although not explicitly related to biological weapons, ProMED shares many goals with Wheelis's surveillance proposal. The program was established in 1993 at a meeting cosponsored by WHO and attended by 60 international experts in human, animal, and plant health. ProMED is working to create a warning system of disease outbreaks "to prevent the spread of 'new' pathogens or the reemergence of known pathogens on an epidemic scale."[36]

A week-long conference in February 1996 moved the initiative closer to action. Like other ProMED leaders, Wheelis believes the system can accomplish the principal aim of his original proposal—to determine whether a disease outbreak is attributable to a hostile source.

Vaccines for Peace

Geissler's Vaccines for Peace proposal came after participants at the 1990 Kühlingsborn conference discussed the danger of biological defense research. Geissler suggested a civilian-controlled immunization program "against hostile use of biological agents and toxins" as a remedy.[37]

Shifting all vaccine work to the civilian sector would "provide the transparency needed to inspire confidence in BWC compliance." Moreover, it could satisfy "the demand for vaccines, including military demand, without raising suspicions regarding offensive intentions."[38]

The goal was unambiguous:

> The transparency-providing and confidence-building effect of Vaccines for Peace would be seriously restricted if it did not lead to a complete shift of *all* [original emphasis] biomedical research and development from military institutions and their contractors to civilian control.[39]

Although subsequent statements on behalf of the program were softened to "conversion, where appropriate,"[40] questions began to mount. Graham Pearson criticized the proposed program on three counts. First, developing vaccines for both public health and biological defense "may result in neither goal and also may unintentionally aid proliferation." Second, if internationally available, the vaccines would allow "a potential aggressor to examine, evaluate, and circumvent, [which] may negate any defense value." Third, converting facilities from military to civilian control will "degrade biological defense capabilities and thus increase the range of materials available and the utility of biological warfare to a potential aggressor."[41]

Geissler rejected Pearson's claim that a vaccine program for both public health and biological defense may meet neither goal. The eradication of smallpox through a worldwide vaccination program proved the opposite, he said.

Geissler disagreed as well that Vaccines for Peace "might enhance rather than reduce the danger of proliferation." He noted that much information about biological weaponry is already publicly available. Moreover, not just the armed forces but all people have a right to protection against biological warfare agents. Putative agents may also be responsible for natural outbreaks of disease.[42]

Geissler's spirited advocacy of Vaccines for Peace was joined by others. They emphasized that any military facility that produced vaccines could easily make offensive agents. Moreover, worldwide availability of vaccines would presumably reduce interest in biological weapons programs.[43]

But Pearson was not alone either. Others began arguing even more vehemently than he that Vaccines for Peace might promote proliferation, that it "should be avoided at all costs."[44] With such pointed opposition, Geissler further modified the proposal.

By 1995, Geissler's plan had been crafted into a far more passive form under the title ProCEID (Program for Controlling Emerging Infectious Diseases). ProCEID would "not request participation and thereby conversion of all former BW facilities, although it may contribute to such conversion in both the short and long term."[45]

The ProCEID steering committee, which included Geissler, had made a political choice. In recognizing that the United States and other countries were unlikely to give up military vaccine research, the new version reflected what was possible. By softening the most controversial idea, the program became broadly palatable. But the core of Geissler's initial proposal was gone.

Geissler was disappointed. "Yes, I am sorry to confess that I had to give up the original Vaccines for Peace idea because it was unrealistic from the viewpoint of the NATO countries, especially the U.S., U.K., and Germany." He felt especially frustrated in view of new revelations about Iraq's biological warfare program. Iraqi activities had eluded UN inspectors for years after the Gulf War. To Geissler this meant "classical attempts to prevent biological war will fail and that the only way out is an international cooperation under complete transparency."[46]

People in quest of biological peace struggle in a sea of crosscurrents. Whatever their favored approach, they face formidable technical, financial, and political obstacles. Enthusiasts of strong defense programs, like Busbee and Pearson, must realize that, in many realistic scenarios, biological defense is impossible. The notion of effective protection for a large population—through vaccinations, outerwear, detection, and identification—seems born more of hope than justifiable expectation.

The U.S. citizenry is no better protected from a biological attack now than when the army's research program began half a century ago.[47] Hundreds of millions of dollars have been spent in the name of biological defense. Additional millions doubtless will be spent. But improvements are not likely to be more than marginal.

Moreover, the questionable safety and ethics of past experiments continue to shadow today's Biological Defense Research Program. For much of the public, *any* biological activity sponsored by the military remains suspect.

On the other hand, advocates of total civilian control of vaccine research, like Geissler and Sidel, are destined for frustration. They face a nearly insurmountable political problem. National attitudes will not permit relinquishment of military responsibility for defense. However noble the intention of a "demilitarized" biological defense program, for many people the phrase sounds like an oxymoron.

Uncompromising persistence can mean less influence on, even irrelevance to, the policy debate. Geissler has ruefully recognized this, as have others who previously were forceful advocates of Vaccines for Peace.

But even if modified, their advocacies can be useful in the service of the larger purpose—to maintain biological peace. Structures, rules, treaties, and programs are at their best when contributing to a moral climate. How important is the moral component in preventing biological war? As argued in the concluding chapter, on the basis of history and experience it can be instrumental. Commonly ignored, sometimes belittled, the moral element deserves nurturing. An overriding sense of impropriety about biological warfare—its repugnance to the conscience of mankind—should be cultivated and reemphasized.

CHAPTER 11

MORALITY, REPUGNANCE, AND BIOLOGICAL WARFARE

The topic was grim. In 1989, a U.S. Senate committee was pondering the spread of biological and chemical weapons. Typically, none of the experts who gave testimony puzzled over the moral implications. But several senators did. Joseph Lieberman worried "that we are crossing another moral threshold, quietly losing our capacity for outrage and action." William Roth wondered "how we develop this moral persuasion again—not to use chemical weapons." From John Glenn: "The moral aspects of this thing to me are—I will not say they are completely figments of somebody's imagination in the modern world, but it approaches that."[1]

Throughout this book, morality has been mentioned in the context of biological and chemical warfare. Although the importance of moral constraints has been stressed, however, the term itself has not been examined. As TV talk shows coarsely demonstrate every day, one person's perversion may be another's persuasion. But moral behavior surely involves more than individual preference.

Morality, of course, means doing the right thing, behaving properly. But who knows what is right and proper? Some rely on scriptural guidance, others on philosophical contemplation, on instinct, or on common sense. In every society, however, there is a code of appropriate behavior, a moral system.

Particular behaviors are viewed as immoral in certain societies, though not in others. Is morality then a cultural construct that varies in different populations? Or does it include at least some universal values shared by people in all societies?

Whatever the ultimate answers, some behaviors have been disparaged in most cultures throughout history. Cannibalism, incest, and homicide have been scorned since ancient times almost everywhere. So have such acts of treachery as the use of poisons in warfare. Such behavior is seen as inherently sneaky, unfair, abhorrent.

The Poison Taboo

In identical words, the Biological and Chemical Weapons Conventions prohibit these weapons "for the sake of all mankind"; and the biological treaty describes the use of germ weaponry as "repugnant to the conscience of mankind." Such descriptions have roots that reach back thousands of years. (Not until the nineteenth century were microorganisms understood to be the cause of infection. Before then, poison and disease were commonly seen as the same. Indeed, the Latin word for poison is *virus*.) Among prohibitions in many civilizations were the poisoning of food and wells and the use of poison weapons. The Greeks and Romans condemned the use of poison in war as a violation of *ius gentium,* the law of nations.[2] Poisons and other weapons considered inhumane were forbidden by the Manu Law of India about 500 B.C. and by the Saracens a thousand years later.[3]

Prohibitions against poisons in war continued through the Middle Ages. In the Western world, their use was considered unchivalrous. A history of the kings of England by the twelfth-century monk William of Malmesbury proclaimed that when someone "uses poisoned arrows, venom, and not valour, inflicts death on the man he strikes. Whatever he effects, then, I attribute to fortune, not courage, because he wars by flight and poison."[4] The prohibitions were reiterated by Hugo Grotius in his 1625 opus on *The Law of War and Peace* and were maintained during the harsh religious conflicts of the time.[5] Emerich de Vattel's eighteenth-century work, *The Law of Nations,* echoed Grotius's condemnation. Poison warfare, wrote the eminent Swiss jurist, is "contrary to the laws of war, and equally condemned by the law of nature and the consent of all civilized nations."[6]

The eighteenth-century English jurist Robert Ward castigated the use of poison arms and the poisoning of wells as a "defalcation of proper principles." Under the law of nations, he wrote, "nothing is more expressly forbidden than the use of *poisoned* arms" [original emphasis].[7]

Contemporary historian John Moon contends that growing nationalism in the eighteenth century weakened the age-old disinclinations about poison weapons. He describes the nationalization of ethics as meaning that nations could use any means to attain their aims in warfare. Military necessity could now be seen as displacing moral considerations.[8]

In the mid-nineteenth century, a few military leaders proposed that toxic weapons be employed, though none actually were. A code of behavior prepared for the U.S. Army during the Civil War reaffirmed the prevailing international attitude.

> The use of poison in any manner, be it to poison wells, or food, or arms, is wholly excluded from modern warfare. He that uses it puts himself out of the pale of law and usages of war.[9]

The 1874 Brussels Declaration barred the use of poisons in war and was followed by the 1899 Hague Conference to codify prohibition through international agreement. The renunciation was repeated at another Hague Conference in 1907. Nevertheless, gas was used in World War I, which may well have reflected a nationalization of ethics and technological advances.

But the experience of large-scale chemical warfare was so horrifying that it led to the 1925 Geneva Protocol, which forbids the use of chemical *and* bacteriological agents in war. Images of victims gasping, frothing, and choking to death had an enormous effect. The text of the protocol expresses the sense of abhorrence felt throughout the world. Appealing to "the conscience and the practice of nations," it affirmed that "the use in war of asphyxiating, poisonous or other gases, and of all analogous liquids, materials or devices, has been justly condemned by the general opinion of the civilized world."[10]

In the end, Moon sees this sense of repugnance as mystical, unknowable. Its causes are "deep and ultimately mysterious," he writes. But he attaches great importance to moral opprobrium. "Ultimately, this deep and lasting revulsion may provide the soundest foundation for the permanent abolition of chemical and biological weapons."[11]

This view about aversion mirrors that of Julian Perry Robinson. A British chemist, Robinson has written about chemical and biological warfare for more than 30 years. Is there not, he rhetorically asks,

> a perception widespread throughout different civilizations that fighting with poison is somehow reprehensible, immoral, utterly wrong—that to resort to chemical warfare is to violate a taboo of a particularly deep kind? For how else, without postulating some such fundamental aversion, can one explain why even today, half a century into the age of nuclear weapons and the moral turpitude which their deterrence strategies have brought, chemical warfare is still regarded as illegitimate?[12]

Political scientist Michael Mandelbaum approaches the question through a comparison of attitudes about nuclear weapons and chemical weapons. Why are people apparently more averse to chemicals, he asks. Because nuclear war is potentially far more destructive, this seems a paradox. But he too ascribes the difference to deep-seated taboos about toxic materials. "The poison taboo recurs through time and across cultures." Moreover, aversion to harmful chemicals is accentuated by contemporary experiences with pollution from pesticides, automobiles, and industry.[13]

Mandelbaum embraces an explanation grounded in genetics. Consonant with evolutionary biology, an inbred aversion to toxic substances could improve a person's chances to survive and reproduce. However, because these inclinations can be overridden—as they assuredly were during World War I—rules and other institutional restraints are important buttresses.[14]

Like Robinson, Mandelbaum focuses on chemicals. But given the century-old understanding that microorganisms can cause disease, all the more does aversion now seem applicable to biological agents. If poisoning with chemicals is viewed with contempt, deliberately infecting people with lethal bacteria is plain ghastly. Thus the passage in the Biological Weapons Convention that uniquely describes biological weapons as "repugnant to the conscience of mankind."

The Army's Problem with Repugnance

Even when the U.S. chemical and biological arms programs were most vigorous, during the 1950s and 1960s, the army continually faced public distaste for these weapons, especially biologicals. In 1954, for example, the army tried to find a civilian firm to manage the biological warfare program. The intention was to tap "the technological talent of American industry." But no appropriate companies were willing. The program was reluctantly continued "as a Chemical Corps operation with government personnel."[15]

Not that outside contractors were unavailable for specific projects. Companies and universities accepted contracts for military biological work then, and many do today. But the army has never been able to escape a stigma about this kind of warfare, which sometimes translated into programmatic obstacles.

In 1955, the army was having "extreme difficulty in filling positions . . . at Camp Detrick," the Maryland site of its biological warfare laboratories. Military spokesmen lamented the inability to attract biological scientists and promised greater effort in recruitment.[16]

The next year, the army called for increased public relations efforts. It sought "recognition of the proper role of chemical and biological warfare" and an end to "a 'hush-hush' policy" on the subject. It wanted to reverse the thinking of civilian scientists who believed the aim of the program was "destructive rather than constructive."[17]

A 1960 report lists recommendations by a committee of the House of Representatives to enhance the biological and chemical warfare program. Among them are "an urgent need for greater public understanding" and "a higher level of support." Implicitly acknowledging distaste for these weapons, the words seem apologetic: "The committee cannot bring itself to describe any weapons of war as 'humane,' and makes no moral judgment" about chemical or biological agents.[18]

A 1962 army document again underscores the struggle to broaden "official and public acceptance of these weapons." A telling entry mentions a contract with a national drug company to develop vaccines for biological agents. The company was not identified because it "demanded and received assurances that the firm name would not be made public." The document explained the company's demand as a reaction to "the equivocal position [given to biological warfare issues] in the public mind." The company was making vaccines, not weapons. Yet even this was enough to worry the firm that its image would be tainted.[19]

The theme was repeated throughout the decade. A 1963 article by the commander of medical research at Fort Detrick expressed frustration that biological weapons appeared "particularly repugnant to many individuals." Recruiting physicians was difficult because they felt that "ethics prohibit their participation" in biological defense research.[20]

In 1967, a writer for *Science* magazine noted:

The chemical and biological weapons program is one of the most secret of all U.S. military efforts—not because it is the most important of our military R&D activities, but because the Pentagon believes it is the most easily misunderstood and because it provokes the most emotional distress and moral turbulence.[21]

The writer's interpretation and that of the army are diametrically opposed. The writer believed that the army maintained secrecy because it feared the public would not respond favorably to more knowledge about its program. But the army felt that it was speaking frankly; the public was just not responding sensibly.

Thus, throughout the period of the U.S. offensive biological warfare program, the army was self-conscious about its work. It continually strained to enhance its image. The attitude seems no less common among people who run today's biological defense program. As recounted in Chapter 3, officials at Fort Detrick are convinced that public distrust can be overcome by better marketing.

But the struggle for public support is not merely about the efficacy of marketing. In addition to distrust because of the army's dark activities in the past, the issue concerns a stigma whose trail is ancient. A public relations effort to reverse deeply felt values about health and morality faces unavoidable frustration.

Repugnance of Biological Agents

Why are biological weapons considered more repugnant than other methods of combat? In part because, like radiation and certain chemicals, they can kill large populations indiscriminately. Combatants and noncombatants alike could suffer from lethal exposures against which they had no chance to defend themselves and whose source they never knew. How unfair, how immoral. But also because, unlike any other weapon, biological agents are an ever-present danger from *natural* sources.

True, under some circumstances, naturally occurring radiation or toxic chemicals may be hazardous. Radioactivity may be dangerously high near uranium concentrations; toxic fumes can spew out of erupting volcanoes. But the number of people affected is limited.

In contrast, everyone experiences bacterial and viral infections. People catch colds and develop illnesses all the time. Among the 2 billion children who contract acute respiratory infections every year, 4 million die as a consequence.[22]

Whether debilitation from infection is slight or severe, every person has an intimate sense of the power of unfriendly germs. Vaccines and antibiotics can prevent and cure many diseases. But not all pathogens respond to treatment. People die continuously everywhere on earth from bacterial and viral infections. Military fatalities in World War I may have reached 20 million, but another 20 million died in the 1918–19 influenza pandemic.[23]

The words we use to describe our relation to disease are common in any military training manual. Bacteria "attack." We are "fighting" a "war" on cancer. Germs are the "enemy." A virus "resists" treatment. The body is "invaded" by microorganisms. A virus "overwhelms" our "defenses." "Humanity's ancient enemies are, after all, microbes," writes Laurie Garrett, whose military allusions can scarcely be less subtle: "In the microbial world warfare is constant."[24]

This language speaks to a deeply imbedded human inclination: we are attuned to fighting *against* germs, not *with* them. To disturb this primal formula is to invite psychological dissonance. This human sensibility has only been accentuated in recent years as previously unknown diseases have threatened large populations. AIDS was unknown before the 1980s. Yet, by the year 2000, the disease may have killed as many as 25 million people.[25]

In the 1990s, people first began to hear about "emerging viruses," a term introduced by Stephen Morse to describe HIV and other new or rapidly expanding viruses in the human population. Ebola virus, dengue virus, hanta virus. Strange and exotic sounding, they seemingly came from nowhere to cause deadly outbreaks. As Morse says, apart from their physical effects the apparently random nature of their appearance is psychologically devastating as well.[26]

Moreover, healing is regarded a virtue in all societies, whether by priest, medicine man, or physician. People invest great moral and material capital in securing health. In the Western world, organized concern about health care predated the modern state. During the Middle Ages, workingmen's associations and guilds had "sickness funds" for their members.[27]

Almost all countries now contribute to their citizens' health care. In 1990, about 10 percent of health-service money spent in developed countries came from central governments. Even among the least developed countries, between 5 and 10 percent of health-care expenditures typically was provided by the governments.[28] In citing the value placed on health universally, Konrad Lorenz noted, "The sanctity of the Red Cross is about the only one of the laws of nations that has always been more or less respected by all nations."[29]

Fostering the Moral Sense

Not surprising, then, that people find efforts purposely to infect others repugnant. Does this mean that what has happened through much of history will continue to happen—that a sense of abhorrence will inhibit the use of poison weapons? Not necessarily.

Some people in any population will feel no moral restraint about infecting others. Aum Shinrikyo, the Japanese cult that released nerve gas in the Tokyo subways in 1995, was also developing biological weapons. But this is only one example of behavior that can cross moral boundaries in any society.

People commit homicide, incest, and other abhorrent acts despite laws and punishments that discourage them. But it is also true that deviant behavior is curbed by formal codes that reinforce the moral sense. And this is the justification for international rules, norms, and punishments that discourage biological and chemical warfare.

Another value of codified prohibitions about such warfare relates to the potential to alter people's values. Persons may be induced to behave in ways to which they are not naturally inclined. Dave Grossman, an army officer and psychologist, provides a compelling example. Citing accounts of military historians, he notes that throughout history most combatants were reluctant to kill their enemies. As recently as World War II, some 85 percent of soldiers fired over the enemies' heads or did not fire at all. This he attributes to "a powerful innate human resistance toward killing one's own species."[30]

How, then, did millions die in the war? From a small minority who did aim to kill, from disease, and from devastation by distant weaponry. Those who fired cannons or released bombs from airplanes could not see their victims. They shared little of the infantryman's reluctance to shoot.

By the time of the Vietnam War, the military had developed techniques to overcome the natural inhibition to kill at close range. Instead of the bull's-eye targets that had been used in earlier instruction, Vietnam trainees shot at lifelike figures. On some occasions, balloon-filled uniforms moved across the kill zone. When hit by bullets, they collapsed. Gunners were repeatedly conditioned to see their targets fall.

Successful shooters received positive reinforcement in the form of praise, recognition, and three-day passes. They were being subjected to "the single most powerful and reliable behavior modification process yet discovered by the field of psychology, and now applied to the field of warfare: operant conditioning." The result: During close combat in Vietnam, only about 5 percent of U.S. soldiers failed to aim to kill.[31]

Although a military man, Grossman is uncomfortable with conditioning that can overcome a healthy repugnance.

> We may never understand the nature of this force in man
> that causes him to strongly resist killing his fellow man, but
> we can give praise to whatever force we hold responsible for
> our existence. And although military leaders responsible
> for winning a war may be distressed by it, as a race we can
> view it with pride.[32]

To preserve this precious force, Grossman warns against trying to weaken our natural reluctance to kill other human beings.[33]

When Daniel Patrick Moynihan wrote an article in 1993 titled "Defining Deviance Down," he was not thinking about biological and chemical warfare. Rather, he focused on behaviors newly redefined in society as "normal" although abnormal by earlier standards. They included novel interpretations of mental health, "alternative" family structures, and a growing acceptance of violent crime. "We are getting used to a lot of behavior that is not good for us," he wrote.[34]

So too can conditioning and habituation affect people with respect to biological weapons. Moynihan's dictum about individual behavior can also apply to groups and to nations. If the number of biological weapons programs throughout the world continues to grow, biological arsenals may come to be seen as normal. The moral sense of repugnance about these weapons could become compromised, as when Iraq went relatively uncensured during its extended use of chemicals against Iran.

One key to deterring use is unavailability. Leon Kass, a physician-ethicist, makes the point in a different context. He abhors the idea of selling human organs. His "untutored repugnance" arises from the nearly universal taboos against cannibalism and other defilements of the human body. Yet he concedes that for his own child he would probably spare no expense to obtain a life-saving kidney.[35]

If organs were not for sale, however, none could be bought. Similarly, if biological and chemical weapons are unavailable—if they have not been produced, stockpiled, and prepared for delivery—questions about usage become moot.

But if such weapons are in stock, even the most ethically sensitive nation might use them as a last resort. A desperate parent might buy an organ for a child; a nation facing devastation might use repugnant weaponry if available.

Protection of a large population against chemical weapons is difficult, against biological weapons virtually impossible. Moral deterrence against their use therefore takes on heightened importance. Even defense research in this area can be problematic for the home population, as described in the first part of this book. In this light, an international norm that does not allow for such weapons is to be cherished.

Banning biological and chemical weapons is easier to accomplish than banning other weapons because biological and chemical weapons provoke an unusual sense of repugnance. Contrary to analyses that commonly ignore or belittle the phenomenon, this natural feeling of antipathy should be appreciated and exploited.

In the 1980s, international responses to Iraq's use of chemicals against Iran were muted or ineffective. To its credit, since the Gulf War much of the international community, through the UN Security Council, has pressed Iraq about its unconventional weapons programs by maintaining sanctions. But even now, UN reports are commonly dry recitations. Expressions of outrage are rare. Any country or group that develops these weapons deserves forceful, unrestrained condemnation.

Words of outrage alone, obviously, are not enough. Enhanced intelligence is important. Carefully controlled defense programs (and an awareness of their limitations) also are appropriate. As mentioned earlier, the public would feel more confident about the ethics and safety of the Biological Defense Research Program if civilians participated in oversight and could confirm the army's claims.

Beyond intelligence and defense, institutions that reinforce good habits and values are essential. Accordingly, treaties, verification, punishment of violators—all have been discussed and endorsed in this book.

The highest priority of the moment in this regard is implementation of the 1993 Chemical Weapons Convention, which outlaws the possession of chemical weapons. The treaty entered into force on April 29, 1997, six months after the 65th ratification. (By the end of 1996, 160 states had signed and 67 had deposited instruments of ratification.) The treaty provides benefits to states parties through information exchange and commercial privileges that are added inducements to join.

The convention lists chemicals that must be declared by the parties. By allowing for short-notice inspections to confirm compliance, temptations to violate the convention are further discouraged. Moreover, its implementation will add momentum to current negotiations to strengthen the 1972 Biological Weapons Convention, which now lacks verification provisions.

Failure to implement the Chemical Weapons Convention successfully will dampen prospects for a verification regime for the biological treaty as well. The most likely consequence would be the continued proliferation of chemical and biological arsenals throughout the world. The longer these weapons are around, the more their sense of illegitimacy erodes, and the more likely they will be used—by armies and by terrorists.

Beside steps to strengthen the biological treaty, actions on another track should help deter biological warfare. Efforts to create worldwide systems to identify and control emerging diseases are gaining attention. ProCEID (Program for Controlling Emerging Infectious Diseases) and ProMED (Program to Monitor Emerging Diseases) were noted earlier. While focusing on disease outbreaks in general, the sponsors of these efforts are sensitive to the possibility of man-made epidemics. Such programs would more likely identify disease outbreaks from hostile sources than is now possible. Especially if implemented in conjunction with a verification regime, they should further discourage nations from contemplating biological warfare.

Such approaches are the muscle of arms control. But their effectiveness ultimately depends on the moral backbone that supports them and the will to use force on its behalf. By underscoring the moral sense behind the formal exclusion of biological and chemical weapons, sustaining their prohibition becomes more likely.

The Passover story depicts the ten plagues as God's imposition on errant people. The eleventh plague is entirely a human invention, a human imposition. Its occurrence represents a moral failure. Its avoidance is a statement of human decency, an act of will born of high principle.

We can hardly afford less.

CHAPTER 1

1. The fifth plague, murrain, does refer to fatal cattle disease (perhaps anthrax). My reference to the eleventh plague as biological and chemical warfare is, of course, allegorical.

2. James L. McWilliams and R. James Steel, *Gas! The Battle for Ypres, 1915* (St. Catherine, Ont.: Vanwell, 1985), 45–49.

3. Stockholm International Peace Research Institute (SIPRI), *The Problem of Chemical and Biological Warfare,* vol. 1, *The Rise of CB Weapons* (New York: Humanities Press, 1971), 129.

4. "Suspect Caught," *Record* (Hackensack, NJ), March 21, 1995, sec. A, p. 9.

5. Nicholas D. Kristof, "Six Killed, Hundreds Hurt As Gas Fills Tokyo Subway," *New York Times,* March 20, 1995, sec. A, p. 1. Although the headline said six killed, the final count was twelve.

6. The descriptions are drawn from Richard Preston, *The Hot Zone* (New York: Random House, 1994), 72–75; Laurie Garrett, *The Coming Plague: Newly Emerging Diseases in a World Out of Balance* (New York: Farrar, Straus and Giroux, 1994), 144, 218–19; and Robert Howard, a spokesperson for the Centers for Disease Control and Prevention, interview, January 11, 1996.

7. Staff statement, U.S. Senate Permanent Subcommittee on Investigations (Minority Staff), *Global Proliferation of Weapons of Mass Destruction, A Case Study on the Aum Shinrikyo: Hearing Before the Permanent Subcommittee on Investigations,* 104th Cong., 1st sess., October 31, 1995, 44.

8. Michael Janofsky, "Looking for Motives in the Plague Case," *New York Times,* May 28, 1995, p. 18; "A Guilty Plea in Mail-Order Bacteria Case," ibid., November 24, 1995, sec. A, p. 23.

9. Kathleen Bailey, interviews, July 13 and 31, 1995.

10. *Convention on the Prohibition of the Development, Production and Stockpiling of Bacteriological (Biological) and Toxin Weapons and on Their Destruction,* opened for signature at London, Moscow, and Washington, DC: April 10, 1972; entered into force: March 26, 1975.

Approximate numbers of countries possessing chemical weapons are reviewed in Gordon M. Burck and Charles C. Flowerree, *International Handbook on Chemical Weapons Proliferation* (New York: Greenwood, 1991), 162–77. Julian Perry Robinson noted that in 1972 the United States suspected four countries of having offensive biological programs. The number presumably fell after the Biological Weapons Convention was signed that year. Robinson listed 22 countries that were publicly named by other states in the 1980s as allegedly having offensive programs. But he cautioned that the accuracy of the allegations "is anyone's guess." J. P. Perry Robinson, "The 1972 Biological Weapons Convention: Aspects of Adherence and Effectiveness," speaking notes for the Dutch government's BWC seminar, Noordwijk, February 19, 1991.

11. John D. Holum, "The Clinton Administration and the Chemical Weapons Convention: Need for Early Ratification," in *Ratifying the Chemical Weapons Convention*, ed. Brad Roberts (Washington, DC: Center for Strategic and International Studies, 1994), 1.

12. Fifteen countries were named at least once in six sources cited in U.S. Congress, Office of Technology Assessment (OTA), *Proliferation of Weapons of Mass Destruction* (Washington, DC: Government Printing Office, August 1993), 82. They were Libya, North Korea, Iraq, Taiwan, Syria, Soviet Union, Israel, Iran, China, Egypt, Vietnam, Laos, Cuba, Bulgaria, and India. Two additional countries, South Korea and South Africa, were named at a 1995 Senate hearing. Statement by Milton Leitenberg, U.S. Senate Permanent Subcommittee on Investigations, *Global Proliferation of Weapons of Mass Destruction: Hearing Before the Permanent Subcommittee on Investigations*, 104th Cong., 1st sess., November 1, 1995, Table 1.

13. U.S. Department of Defense, "Nuclear/Biological/Chemical (NBC) Warfare Defense," annual report to Congress, June 1994, 3.

14. Graham S. Pearson, "Vaccines for Biological Defence: Defence Considerations," in *Control of Dual-Threat Agents: The Vaccines for Peace Program*, ed. Erhard Geissler and John P. Woodall, Stockholm International Peace Research Institute (SIPRI) (New York: Oxford University Press, 1994), 154–55.

15. Leonard A. Cole, *Clouds of Secrecy: The Army's Germ Warfare Tests Over Populated Areas* (Savage, MD: Rowman and Littlefield, 1988, 1990), 23–31. The spores had largely settled underground and were not

likely to become airborne again, which has led some to think the danger may have been minimal.

16. Robert Harris and Jeremy Paxman, *A Higher Form of Killing: The Secret Story of Chemical and Biological Warfare* (New York: Hill and Wang, 1982), 105. Pearson disagrees with Watson, contending that the city could have been decontaminated earlier. Interview, September 21, 1994.

17. Brad Roberts, "New Challenges and New Policy Priorities for the 1990s," in *Biological Weapons: Weapons of the Future?*, ed. Brad Roberts (Washington, DC: Center for Strategic and International Studies, 1993), 94–95.

18. J. H. Rothschild, Brigadier General, U.S.A. (Ret.), *Tomorrow's Weapons: Chemical and Biological* (New York: McGraw-Hill, 1964), 22.

19. OTA, *Proliferation of Weapons of Mass Destruction*, 54.

20. Neil C. Livingstone and Joseph D. Douglass, Jr., *CBW: The Poor Man's Atomic Bomb* (Cambridge, MA: Institute for Foreign Policy Analysis, 1984), 7.

21. *Protocol for the Prohibition of the Use in War of Asphyxiating, Poisonous or Other Gases, and of Bacteriological Methods of Warfare,* signed at Geneva: June 17, 1925.

22. John W. Powell, "A Hidden Chapter in History," *Bulletin of the Atomic Scientists* 37, no. 8 (October 1981), 45–49; Erhard Geissler, ed., *Biological and Toxin Weapons Today,* SIPRI (New York: Oxford University Press, 1986), 10.

23. At times during World War II, Allied commanders worried that the Germans and Japanese might use chemical weapons. When planning the 1944 Normandy invasion in particular, they "speculated on the probability" of a German gas attack. SIPRI, *The Problem of Chemical and Biological Warfare,* vol. 1, 297. But the public had no sense of imminence, because the planning and speculation were secret.

24. In World War II, President Roosevelt declared that the United States would not be the first to use chemical or biological weapons. Under President Eisenhower in 1956, the policy was altered to allow for presidential discretion. The U.S. use of irritant agents and herbicides in the Vietnam War was an ambiguous case. The action prompted a 1969 UN General Assembly resolution condemning the use of chemical or biological agents that have a "direct toxic

effect on man, animals or plants." Eighty nations supported the resolution, and the three that opposed included the United States. Many of the 36 that abstained were viewed as favorable to the resolution but did not wish to desert a major ally. SIPRI, *The Problem of Chemical and Biological Warfare*, vol. 5, *The Prevention of CBW*, 41, 53.

25. Remarks of President Nixon on announcing the chemical and biological defense policies and programs, Office of the White House Press Secretary, November 24, 1969; Marie Isabelle Chevrier, "Verifying the Unverifiable: Lessons from the Biological Weapons Convention," *Politics and the Life Sciences* 9, no. 1 (August 1990), 98.

26. *Report of the Chemical Warfare Review Commission* (Washington, DC: Government Printing Office, June 1985), xvi, 69.

27. U.S. Senate Subcommittee on Oversight of Government Management of the Committee on Governmental Affairs, *Department of Defense Safety Programs for Chemical and Biological Warfare Research: Hearings Before the Subcommittee on Oversight of Government Management*, 100th Cong., 2d sess., July 27–28, 1988, 3.

28. *Biological Weapons Convention*, Articles I and IV.

29. *Convention on the Prohibition of the Development, Production, Stockpiling and Use of Chemical Weapons and on Their Destruction*, opened for signature at Paris, January 13, 1993.

30. Susan Wright, ed., *Preventing a Biological Arms Race* (Cambridge, MA: MIT Press, 1990), 5.

31. Joseph D. Douglass, Jr., and Neil C. Livingstone, *America the Vulnerable: The Threat of Chemical and Biological Warfare* (Lexington, MA: Lexington Books, 1987), 177.

32. "$16.5 Million Budgeted for Dugway Germ Lab," *Deseret News* (Salt Lake City), May 12, 1993, sec. B, p. 1.

33. In addition to previous citations—Cole, *Clouds of Secrecy;* Geissler, *Biological and Toxin Weapons Today;* Geissler and Woodall, *Control of Dual-Threat Agents;* Roberts, *Biological Weapons;* Wright, *Preventing a Biological Arms Race*—other books limited to biological warfare issues include W. Seth Carus, *"The Poor Man's Atomic Bomb?" Biological Weapons in the Middle East* (Washington, DC: Washington Institute for Near East Policy, 1991); Malcolm Dando, *Biological*

Warfare in the 21st Century (London: Brassey's, 1994); Erhard Geissler, ed., *Strengthening the Biological Weapons Convention by Confidence-Building Measures,* SIPRI (New York: Oxford University Press, 1990); Erhard Geissler and Robert H. Haynes, eds., *Prevention of a Biological and Toxin Arms Race and the Responsibility of Scientists* (Berlin: Akademie-Verlag, 1991); Sheldon H. Harris, *Factories of Death: Japanese Biological Warfare 1932–45 and the American Cover-Up* (New York: Routledge, 1994); S. J. Lundin, ed., *Views on Possible Verification Measures for the Biological Weapons Convention,* SIPRI (New York: Oxford University Press, 1991); Jeanne McDermott, *The Killing Winds: The Menace of Biological Warfare* (New York: Arbor House, 1987); Charles Piller and Keith Yamamoto, *Gene Wars: Military Control over the New Genetic Technologies* (New York: Beech Tree Books, 1988); Sterling Seagrave, *Yellow Rain: A Journey Through the Terror of Chemical Warfare* (New York: M. Evans, 1981) [Despite reference to chemical warfare in the title, yellow rain ostensibly was a biologically derived toxin made by the Soviets.]; Nicholas A. Sims, *The Diplomacy of Biological Disarmament* (New York: St. Martin's, 1988); Barend ter Haar, *The Future of Biological Weapons* (New York: Praeger, 1991); Peter Williams and David Wallace, *Unit 731: Japan's Secret Biological Warfare in World War II* (New York: Free Press, 1989); Raymond A. Zilinskas, ed., *The Microbiologist and Biological Defense Research: Ethics, Politics, and International Security* (New York: New York Academy of Sciences, 1992).

34. OTA, *Proliferation of Weapons of Mass Destruction,* 61.

35. "The Technology of Biological Arms Control and Disarmament," NATO Advanced Research Workshop, Budapest, March 28–30, 1996. Press release, U.S. Senate Permanent Subcommittee on Investigations of the Committee on Governmental Affairs, "Global Proliferation of Weapons of Mass Destruction: Subcommittee to Hold Series of Hearings in March," March 4, 1996.

36. Wright, 9, 75.

37. Jeffrey W. Almond, quoted in Charles Siebert, "Small Pox Is Dead, Long Live Smallpox," *New York Times Magazine,* August 21, 1994, 55.

38. The White House, news release, February 14, 1970.

39. Roberts himself does not dismiss the value of norms. Brad Roberts, "The Chemical Weapons Convention and World Order," in *Shadows*

and Substance: The Chemical Weapons Convention, ed. Bennoit Morel and Kyle Olson (Boulder, CO: Westview, 1993), 13.

40. Matthew Meselson, "The Chemical and Biological Weapons Conventions" (paper presented at the annual meeting of the American Association for the Advancement of Science, Atlanta, February 17, 1995).

CHAPTER 2

1. Leonard A. Cole, *Clouds of Secrecy: The Army's Germ Warfare Tests Over Populated Areas* (Savage, MD: Rowman and Littlefield, 1990), chaps. 7 and 8.

2. *Barrett v United States,* 660 F. Supp, 1291 (SDNY 1987), 1300. Also, testimony of Elizabeth Barrett, House Committee on Government Operations, *Cold War Era Human Subject Experimentation: Hearings Before the Subcommittee on Legislation and National Security,* 103d Cong., 2d sess., September 28, 1994 (distributed at the hearing).

3. Eric Wick Olson, in testimony to the House Committee, *Cold War Era Human Subject Experimentation,* September 28, 1994. "Eye on America," *CBS Evening News,* November 28, 1994.

4. Tom Meersman, "Safety of 1953 Army Spraying Questioned Now," *Star Tribune* (Minneapolis), June 11, 1994, sec. A, p. 10. Through 1995, former Clinton students Diane Gorney, Richard Meixner, and Joyce Carlson had reached some 350 classmates who provided information about their health histories, according to Gorney. Interview, November 27, 1995.

5. U.S. Senate Committee on Human Resources, *Biological Testing Involving Human Subjects by the Department of Defense, 1977: Hearings Before the Subcommittee on Health and Scientific Research,* 95th Cong., 1st sess., March 8 and May 23, 1977.

6. Ibid., 19.

7. Cole, 44–58.

8. Much of the material in this chapter is based on information from annual reports produced by the Army Chemical Corps after World War II. They were obtained in 1994 in response to a Freedom of Information Act request. The reports summarize the chemical, biological, and radiological activities that fell under the Corps' authority. The earliest were called *Annual Report* or *Summary History,* but beginning in 1953 each carried the suggestive title *Summary of Major Events and Problems.* The authors are not identified by name, but the publishing source is the U.S. Army Chemical Corps Historical Office. The first of the available volumes was *Annual Report of*

the Chemical Warfare Service for the Fiscal Year 1945, followed by *Summary History of the Chemical Corps, 25 June 1950–8 September 1951* and *Summary History of Chemical Corps Activities, 9 September 1951 to 31 December 1952.* Thereafter, annual volumes were titled: United States Army Chemical Corps, *Summary of Major Events and Problems,* U.S. Army Chemical Corps Historical Office, Army Chemical Center, MD, for fiscal years 1953 through 1962.

9. *Summary,* fiscal year 1958, 108–109.

10. I am grateful to Julian Perry Robinson for pointing this out in a personal communication, January 29, 1996.

11. *Summary,* fiscal year 1958, 109.

12. Ibid., 110.

13. Herbert E. Beningson, *The Development and Operation of the LAC Disseminator,* BWL Technical Memorandum 2-26, Biological Warfare Laboratories, Fort Detrick, Frederick, MD, September 1958, 3–5.

14. Ibid., 9.

15. Ibid., app. A, 4.

16. Ibid., 5.

17. *Biological Testing Involving Human Subjects,* 109.

18. Personal communication from Colonel James D. Tipton, commanding officer, U.S. Army Dugway Proving Ground, Utah, April 20, 1988.

19. News release, Public Affairs Office, U.S. Army Dugway Proving Ground, Utah, c. June 1993.

20. Graham S. Wilson and Ashley Miles, *Topley's and Wilson's Principles of Bacteriology and Immunity,* vol. 1 (Baltimore: Williams & Wilkins, 1975), 1100–1101; Bernard R. Davis, Renato Delbucco, Herman N. Eisen, and Harold S. Ginsberg, *Microbiology,* 3d ed. (Hagerstown, MD: Harper & Row, 1980), 709.

21. Beningson, app. B.

22. *Biological Testing Involving Human Subjects,* 134.

23. Philip A. Leighton, *The Stanford Fluorescent-Particle Tracer Technique: An Operational Manual* (Palo Alto: Stanford University Department of Chemistry, June 1955; 2d printing, June 1958), 6–16.

24. Ibid., 14–15.

25. Ibid.

26. Frank Princi, "A Study of Industrial Exposures to Cadmium," *Journal of Industrial Hygiene and Toxicology* 29, no. 5 (September 1947): 319.

27. Harriet L. Hardy and John B. Skinner, "The Possibility of Chronic Cadmium Poisoning," *Journal of Industrial Hygiene and Toxicology* 29, no. 5 (September 1947): 323.

28. Lars Friberg, "Proteinuria and Kidney Injury among Workmen Exposed to Cadmium and Nickel Dust," *Journal of Industrial Hygiene and Toxicology* 30, no. 1 (January 1948): 36.

29. Frank Princi and Erving Geever, "Prolonged Inhalation of Cadmium," *Archives of Industrial Hygiene and Occupational Medicine* 1 (1950).

30. Leon Prodan, "Cadmium Poisoning: II. Experimental Cadmium Poisoning," *The Journal of Industrial Hygiene* 14, no. 5 (May 1932): 174, 192.

31. Lee Davidson, "Tests Exposed U.S. to Chemical," *Deseret News* (Salt Lake City), April 14, 1991, sec. A, p. 5.

32. L. Arthur Spomer, "Fluorescent Particle Atmospheric Tracer: Toxicity Hazard," *Atmospheric Environment* 7 (1973): 353.

33. Ibid.

34. Major Lee DeLorme, quoted in *Chicago Tribune*, August 4, 1980.

35. U.S. Army Environmental Hygiene Agency, "Assessment of Health Risk, Minneapolis, Minnesota," Health Risk Assessment Study No. 64-50-93QE-94, July 10, 1994, 1.

36. Ibid., 27–29.

37. *Summary*, fiscal year 1959, 101.

38. Ibid., 101–103.

39. Ibid., 103.

40. Ibid.

41. Barry Miller, interview, June 13, 1994.

42. John Woodall, interview, November 11, 1994.

43. *Summary*, fiscal year 1959, 162.

44. *Summary*, fiscal year 1955, 48–49.

45. *Summary*, fiscal year 1957, 97.

46. Ibid., 98.

47. Ibid.

48. Ibid., 99.

49. *Summary*, fiscal year 1958, 100–101.

50. Ibid.

51. Quoted in Department of the Army, Office of the Inspector General and Auditor General, Subject: "Research Report Concerning

the Use of Volunteers in Chemical Agent Research," Washington, DC, July 21, 1975, 40.

52. Ibid., 40–41.

53. "Response to Man in a BW Field Test, Operation 'CD-22', BW 3-55," DPGR 186, Dugway Proving Ground, Utah, December 17, 1956, 7.

54. Ibid., 8.

55. Ibid., 13, 36–37.

56. Ibid., 9–10.

57. Russell L. Cecil and Robert F. Loeb, *A Textbook of Medicine* (Philadelphia: Saunders, 1952), 90–91.

58. "Response to Man in a BW Field Test, Operation 'CD-22', BW 3-55," 21. This spraying, in July 1955, was one of several outdoor tests that were part of Operation CD-22.

59. Ibid., 37–38.

60. "Research Report Concerning the Use of Volunteers in Chemical Agent Research," 40.

61. *Summary*, fiscal year 1959, 107–108.

62. Ibid.

63. Colonel Dan Crozier, "The Threat of Biological Weapons Attack," *Military Medicine* 128 (1963): 84.

64. *Summary*, fiscal year 1959, 108.

65. Cole, *Clouds of Secrecy*, 148–49.

66. U.S. Secretary of Defense, Memorandum for the Secretary of the Army, Secretary of the Navy, Secretary of the Air Force, Subject: Use of Human Volunteers in Experimental Research, Washington, DC, February 26, 1953. This and other memoranda on the subject were provided to me by Stephen Klaidman, director of communications for the Advisory Committee on Human Radiation Experiments. The committee was established by the president in 1994 to look into the ethics of human radiation experiments.

67. Department of the Army, Memorandum for the Chief of Staff, Subject: Use of Volunteers in Research, Washington, DC, November 5, 1953.

68. Ibid.

69. On June 30, 1953, the army chief of staff "published a memorandum . . . for the Chief Chemical Officer and The Surgeon General of the Army which implemented the Secretary of Defense's guidance. This directive was initially classified 'top secret.' It was

regraded 'confidential' and then 'unclassified' in July 1954 at the urging of the Secretary of the Army." "Research Report Concerning the Use of Volunteers in Chemical Agent Research," 77.

70. Ibid., 78.

71. Ibid., 80.

72. "Project Whitecoat," vol. 2, compiled by SP/5 Errol L. Chamness, USAMRIID, Summer 1971.

73. "Research Report Concerning the Use of Volunteers in Chemical Agent Research," 40–41, 80.

74. Ibid., 81.

75. Ibid., 85–87.

CHAPTER 3

1. U.S. Department of the Army, Dugway Proving Ground, "Record of Decision: Construction and Operation of Life Sciences Test Facility," Dugway, Utah, March 1993, 2.

2. U.S. Department of the Army, U.S. Medical Research and Development Command, Final Programmatic Environmental Impact Statement, "Biological Defense Research Program," Fort Detrick, Frederick, MD, April 1989, ES-6.

3. Ibid., app. 8, p. 8.

4. John M. R. Bull and Peter Kelly, "Army Disease Tests Raise Fears," *Patriot-News* (Harrisburg, PA), June 25, 1989, sec. A, p. 6.

5. Lee Davidson, "Army Unsure What Lurks Below Dugway Surface," *Deseret News* (Salt Lake City), March 15, 1988, sec. A, p. 1.

6. U.S. Department of the Army, U.S. Medical Research and Development Command, Draft Programmatic Environmental Impact Statement, "Biological Defense Research Program," Fort Detrick, Frederick, MD, May 1988.

7. Final Programmatic Environmental Impact Statement, chap. 5, p. 9.

8. Ibid., app. 9, p. 56.

9. U.S. Senate, "Report of the Majority Staff of the Senate Subcommittee on Oversight of Government Management on DOD's Safety Programs for Chemical and Biological Research," May 12, 1988; John H. Cushman, Jr., "Perils Seen in Pentagon's Biological Research," *New York Times,* May 12, 1988, sec. A, p. 28.

10. Senator Carl Levin, U.S. Senate Committee on Governmental Affairs, *Department of Defense Safety Programs for Chemical and Biological*

Warfare Research: Hearings Before the Subcommittee on Oversight of Government Management, July 27–28, 1988, 3.

11. U.S. General Accounting Office, "DOD's Risk Assessment and Safeguards Management of Chemical and Biological Warfare Research and Development Facilities," GAO/T-EMD-88-10 (Washington, DC: Government Printing Office, July 1988).

12. U.S. General Accounting Office, "Biological Warfare: Better Controls in DOD's Research Could Prevent Unneeded Expenditures," GAO/NSIAD-91-68 (Washington, DC: General Accounting Office, December 1990).

13. U.S. Department of Labor, Occupational Safety and Health Administration, "Evaluation of the U.S. Department of the Army's Occupational Safety and Health Program for the Chemical and Biological Defense Research Laboratories," February 1991.

14. U.S. Department of Defense, Department of the Army, "Biological Defense Safety Program (Technical Safety Requirements)," DA Pamphlet 385-69, *Federal Register,* vol. 56, no. 18, January 28, 1991, and vol. 56, no. 44, March 6, 1991.

15. William Wortley in telephone inquiry, September 23, 1991. See also "Safety First in Germ Warfare," *Science* 253, no. 5021 (August 16, 1991), 727.

16. Seth Shulman, *Biohazard: How the Pentagon's Biological Warfare Research Program Defeats Its Own Goals* (Washington, DC: Center for Public Integrity, 1993), 31.

17. The subjects were volunteers from the professional staffs of USAMRIID and Walter Reed Army Hospital, according to Levitt in an interview, October 2, 1991.

18. Senate Hearings, *Department of Defense Safety Programs for Chemical and Biological Warfare Research,* 3–4.

19. Ibid., 6.

20. Ibid., 109.

21. Ibid., 110.

22. Final Programmatic Environmental Impact Statement, app. 15, p. 4.

23. Unless otherwise indicated, comments from Levitt are from an interview, April 8, 1994.

24. Letter to Senator Charles McC. Mathias from Colonel Richard Singleton, Office of the Inspector General, Department of the Army, May 13, 1986.

25. Helen Ramsburg, interview, April 27, 1994.

26. Unless otherwise noted, comments from Fort Detrick personnel were made in interviews at the base on May 18, 1994.

27. "Twenty-five Years, USAMRIID, 1969–1994," U.S. Army Medical Research Institute of Infectious Diseases, Fort Detrick, MD, 1994, 9.

28. Leonard A. Cole, *Clouds of Secrecy: The Army's Germ Warfare Tests Over Populated Areas* (Savage, MD: Rowman and Littlefield, 1988, 1990), 32–33.

29. Norman M. Covert, *Cutting Edge, A History of Fort Detrick, Maryland, 1943–1993,* Headquarters U.S. Army Garrison, Public Affairs Office, Fort Detrick, MD, 1993, 48–49.

30. Although building 470 is locked and empty, visitors have been escorted inside without special protective gear. Anthrax spores in cracks or crevices are unlikely to cause problems unless they become airborne, which probably explains why the building has not been renovated or razed.

CHAPTER 4

1. "Dugway Proving Ground History" and "Joint Service Chemical and Biological Testing," Public Affairs Office, U.S. Army Dugway Proving Ground [c. 1994].

2. U.S. Army Dugway Proving Ground, Installation Environmental Assessment, prepared by environmental and ecology staff, Environmental and Life Science Division, January 1982, 7, 15, 18, 64.

3. Lee Davidson, "Like Sheep to the Slaughter?" *Deseret News* (Salt Lake City), May 30, 1993, sec. B, p. 1.

4. Ibid., sec. B, p. 2.

5. Ibid.

6. Testimony by Earl P. Davenport before the U.S. Senate Committee on Veterans' Affairs, May 6, 1994, distributed at the hearing.

7. Ibid.

8. Letter from Attilio D. Renzetti, Jr., M.D., to U.S. Department of Labor, copy to U.S. Army Health Clinic, Dugway Proving Ground, September 10, 1984.

9. Leroy W. Metker, chief, Toxicity Evaluation Branch, Toxicology Division, to Major Weyandt, HHA, April 6, 1984.

10. Toxicity Status and Current Use of Dimethyl Methyl Phosphonate (DMMP), USAEHA, Aberdeen Proving Ground, MD, to HQDA (DASG-PSP), 5111 Leesburg Pike, Falls Church, VA, October 10, 1986.

11. Larry K. Whisenant, chief, Safety Office, Subject: Restriction in the use of DMMP for MCPE testing at DPG, to MT-TM-CB, Attention: Gladden, August 24, 1988.

12. Matthew S. Brown, "Ex-Dugway Worker Blames Simulant for Health Woes," *Deseret News* (Salt Lake City), April 24, 1994, sec. A, p. 7.

13. Davenport's annual medical health reports and related documents were provided to the Senate Committee on Veterans' Affairs in 1994.

14. U.S. Department of Labor, Loren L. Smith, Office of Workers' Compensation Programs, case no. A12-75549, July 20, 1993.

15. U.S. Army Dugway Proving Ground, Public Affairs Office, news release, July 26, 1993.

16. U.S. Army Dugway Proving Ground, Public Affairs Office, news release, May 2, 1991.

17. U.S. Army Dugway Proving Ground, Material Test Directorate, "Biological Defense Update," for governor's Technical Review Committee, Salt Lake City, July 26, 1993.

18. Wolfgang K. Joklik et al., eds., *Zinsser Microbiology*, 20th ed. (Norwalk, CT: Appleton and Lange, 1992), 111.

19. Zell McGee, interview, March 28, 1994.

20. Susan Mottice, interview, March 28, 1994.

21. Kari Sagers, interview, March 28, 1994.

22. "Environmental Assessment for Battledress Overgarment Penetration Study, Phase 1—Field Test, at U.S. Army Dugway Proving Ground, Utah," prepared by James H. Wheeler, Christina M. Wheeler, and John S. Allan, Materiel Test Directorate, Life Sciences Division, Environmental Technology Section, U.S. Army Dugway Proving Ground, Dugway, Utah, July 1993 (mimeographed).

23. Information about the subjects is from Matthew Brown, "Army Defends 'Informed Consent' of Volunteers," *Deseret News* (Salt Lake City), April 24, 1994, sec. A, pp. 1 and 6, and from Brown's conversation with me on March 11, 1994.

24. Another disconcerting incident occurred four months after the test. The army provided a rewritten consent form that listed the chemicals to the Senate Committee on Veterans' Affairs. The new "final" form evidently was prepared after the test. This information was provided to me by committee staff member Patricia Olson on March 29, 1994.

25. Letter from Senator James Sasser to Secretary of Defense Caspar W. Weinberger, October 31, 1984, Washington, DC, Office of Senator James Sasser.

26. Memorandum Opinion and Order by Judge Joyce Hens Green, re. Foundation on Economic Trends, et al., Plaintiffs, vs. Caspar W. Weinberger, et al., Defendants, Civil Action No. 84-3452, U.S. District Court for the District of Columbia, November 21, 1984.

27. Guy Boulton, "Bangerter Opposes 'Level 4' Dugway Germ Lab," *Salt Lake Tribune*, March 25, 1988, sec. B, p. 1.

28. "Oppose New Dugway Lab Until Questions Answered," *Deseret News* (Salt Lake City), March 15–16, 1988, sec. A, p. 12; "Standing Up for Utah," *Salt Lake Tribune*, March 18, 1988, sec. A, p. 14.

29. Colin Norman, "Army Shifts on Dugway Lab," *Science* 241, no. 4874 (September 30, 1988), 1749.

30. U.S. Army Dugway Proving Ground, Public Affairs Office, news release, May 2, 1991.

31. Joseph Bauman, "Panel Studies Planned Germ, Chemical Testing," *Deseret News* (Salt Lake City), May 3, 1991, sec. B, p. 8.

32. Kenneth Buchi, interview, July 31, 1991.

33. Bauman, sec. B, p. 8.

34. Jim Woolf, "Downwinders Sue to Block Renewed Tests at Dugway," *Salt Lake Tribune*, July 3, 1991, sec. B, p. 2.

35. U.S. District Court, District of Utah, Central Division, Motion for Preliminary Injunction, Downwinders, Inc., a Nonprofit Corporation, Plaintiff, vs. Dick Cheney, Secretary of Defense, and Michael P. W. Stone, Secretary of the Army, Defendants, July 1, 1991.

36. Jim Woolf, "Judge Dismisses Lawsuit to Halt Testing," *Salt Lake Tribune*, January 5, 1995, sec. B, p. 3.

37. Joseph Bauman, "U. Hospital, Army Spar over Treating Workers Exposed to Organisms," *Deseret News* (Salt Lake City), June 6, 1991, sec. A, p. 1.

38. Jim Woolf, "Biological Agents? Army Briefing Miffs U. Doctors," *Salt Lake Tribune*, July 18, 1991, sec. B, p. 1.

39. "U Doctors Tour Dugway," *Deseret News* (Salt Lake City), August 8, 1991, sec. B, p. 1.

40. U.S. Army Dugway Proving Ground, Public Affairs Office, news release, September 24, 1991.

41. Nancy Melling, interview, March 28, 1994.

42. Suzanne Winters, interview, January 17, 1995.

43. Jay Jacobson, interview, January 17, 1995.

44. Susan Mottice, interview, January 17, 1995.

45. Kenneth Buchi, interview, April 4, 1994.

46. Minutes, Dugway Technical Review Committee meeting, September 10, 1993, 3.

CHAPTER 5

1. UN Security Council Resolution 678, enacted November 29, 1990, implicitly allowed member states to use force if Iraq did not withdraw from Kuwait by January 15, 1991. The resolution authorized them to "use all necessary means" to implement earlier resolutions calling for Iraqi withdrawal.

2. Most countries in the coalition provided only financial, medical, or other support services. The European Union contributed funds in addition to assistance from individual member states. Participants included countries throughout the world: from Africa—Sierra Leone, Senegal, Niger; from Eastern Europe—Czechoslovakia, Poland, Romania; from Asia—Japan, Pakistan, Singapore, South Korea, Hong Kong. See Elaine Sciolino, *The Outlaw State: Saddam Hussein's Quest for Power and the Gulf Crisis* (New York: Wiley, 1991), 25; and Lawrence Freedman and Efraim Karsh, *The Gulf Conflict, 1990–1991: Diplomacy and War in the New Order* (Princeton, NJ: Princeton University Press, 1993), 342–61.

3. Judith Miller, "Morocco's Call for Arab Meeting on the Gulf Has Diplomats Astir," *New York Times,* November 14, 1990, sec. A, p. 14.

4. John R. MacArthur, *Second Front: Censorship and Propaganda in the Gulf War* (New York: Hill and Wang, 1992), 90.

5. Books about the Gulf War have shown surprisingly little interest in exploring the policies that permitted Iraq to build chemical and biological arsenals. This is true whether the authors criticized the military action against Iraq or supported it. Examples of the former are MacArthur, *Second Front;* Jean Edward Smith, *George Bush's War* (New York: Holt, 1992); and Martin Yant, *Desert Mirage: The True Story of the Gulf War* (Buffalo, NY: Prometheus, 1991). Examples of the latter include Freedman and Karsh, *The Gulf Conflict;* and Molly Moore, *A Woman at War: Storming Kuwait with the U.S. Marines* (New York: Scribner's, 1993). Elaine Sciolino did review reasons the Reagan administration "left office without ever addressing Iraq's use of chemical weapons," though her assessment was only one page long. Sciolino, *The Outlaw State,* 171–72.

6. Maureen Dowd, "Storm's Eye: Bush Decides to Go to War," *New York Times,* January 17, 1991, 16.

7. Nicholas D. Kristof, "Tension Rises in Korean Stare-Down," *New York Times,* January 28, 1996, 10. The article cites a military assessment that, if North Korea were to attack South Korea, there would be a "high likelihood" that North Korea would use chemical weapons.

8. Anthony H. Cordesman and Abraham R. Wagner, *The Lessons of Modern War,* vol. 2, *The Iran-Iraq War* (Boulder, CO: Westview, 1990), 507.

9. Ibid., 507–509.

10. Herbert Krosney, *Deadly Business: Legal Deals and Outlaw Weapons* (New York: Four Walls Eight Windows, 1993), 44–45.

11. Ibid., 48.

12. Ibid., 48–49.

13. The list includes companies that supplied materials for Iraq's nuclear, chemical, and biological programs. Kenneth R. Timmerman, *The Poison Gas Connection,* a special report commissioned by the Simon Wiesenthal Center, Los Angeles, 1990, 46.

14. Ibid., 52.

15. *Response* 11, no. 4 (Winter 1990/1991), 2–4. The bank's role in Iraq's military buildup, including its chemical and biological weapons programs, and the bank's connection to American government and business interests are reviewed in Alan Friedman, *Spider's Web: The Secret History of How the White House Armed Iraq* (New York: Bantam, 1993), 84–147.

16. Thomas C. Wiegele, *The Clandestine Building of Libya's Chemical Weapons Factory: A Study in International Collusion* (Carbondale: Southern Illinois University Press, 1992), 159.

17. Timmerman, 17–18; Krosney, 108–110, 293.

18. S. J. Lundin and Thomas Stock, "Chemical and Biological Warfare: Developments in 1990," *SIPRI Yearbook 1991, World Armaments and Disarmament,* Stockholm International Peace Research Institute (SIPRI) (New York: Oxford University Press, 1991), 106–107.

19. Trevor Wilson, "Chemical Disarmament versus Chemical Nonproliferation," in *Chemical Disarmament and U.S. Security,* ed. Brad Roberts (Boulder, CO: Westview, 1992), 58–59.

20. *New York Times,* February 25, 1989, 6.

21. "Secret Factory Found in Iraq," *Record* (Hackensack, NJ), August 8, 1991, sec. A, p. 17.

22. Barbara Crossette, "Iraq Hinders Arms Monitors, U.N. Panel Reports," *New York Times,* December 21, 1994, sec. A, p. 10.

23. Stephen Engelberg, "Iraq Said to Study Biological Arms," *New York Times,* January 18, 1989, sec. A, p. 7.

24. Deborah Orin and Uri Dan, "Iraq Building Weapons for Germ Warfare," *New York Post,* April 12, 1990, 4.

25. ATCC, *Catalogue of Bacteria and Bacteriophages,* 18th ed. (Rockville, MD: American Type Culture Collection, 1992), ii, iv.

26. U.S. Senate, a report by chairman Donald W. Riegle, Jr., and ranking member Alfonse M. D'Amato of the Committee on Banking, Housing and Urban Affairs, *U.S. Chemical and Biological Warfare-Related Dual Use Exports to Iraq and Their Possible Impact on the Health Consequences of the Persian Gulf War,* May 25, 1994, 39–41.

27. Kevin Merida and John Mintz, "Rockville Firm Shipped Germ Agents to Iraq, Riegle Says," *Washington Post,* February 10, 1994, sec. A, p. 8.

28. Patrick Burke, interview, March 6, 1995.

29. Senate Committee, *U.S. Chemical and Biological Warfare-Related Dual Use Exports to Iraq,* 45–47.

30. *Congressional Record,* 103d Cong., 2d sess., February 9, 1994, S1197.

31. Dilip Hiro, *The Longest War: The Iran-Iraq Military Conflict* (New York: Routledge, 1991), 43. Purges of air force officers in 1981, although short-lived, further weakened Iranian air strength. Cordesman and Wagner, 118.

32. Hiro, 215–25.

33. *Protocol for the Prohibition of the Use in War of Asphyxiating, Poisonous or Other Gases, and of Bacteriological Methods of Warfare,* signed at Geneva: June 17, 1925.

34. Cordesman and Wagner, 514; Krosney, 42.

35. Hiro, 102–105.

36. Bernard Gwertzman, "U.S. Says Iraqis Used Poison Gas Against Iranians in Latest Battles," *New York Times,* March 6, 1984, sec. A, p. 1; Seymour M. Hersh, "U.S. Aides Say Iraqis Made Use of a Nerve Gas," *New York Times,* March 30, 1984, sec. A, p. 1.

37. Richard Bernstein, "U.N. Council Votes to Condemn Iranian Attacks on Gulf Shipping," *New York Times,* June 2, 1984, sec. A, p. 1.

38. Hiro, 222.

39. Ibid., 205.

40. Robin Wright, *In the Name of God: The Khomeini Decade* (New York: Simon and Schuster, 1989), 184.

41. John Bulloch and Harvey Morris, *The Gulf War: Its Origins, History and Consequences* (London: Methuen, 1989), 244.

42. Serge Schmemann, "Iraq Acknowledges Its Use of Gas but Says Iran Introduced It in War," *New York Times,* July 2, 1988, 3.

43. Paul Lewis, "U.N. Chief says He Will Declare a Cease-Fire in the Iran-Iraq War," *New York Times,* August 2, 1988, sec. A, pp. 1, 9.

44. Bulloch and Morris, 259.

45. *New York Times,* August 27, 1988, 3.

46. Bulloch and Morris, xx–xi.

47. U.S. Senate Committee on Foreign Relations, *Chemical and Biological Weapons Threat—The Urgent Need for Remedies: Hearings Before the Committee on Foreign Relations,* 101st Cong., 1st sess., January 24, March 1, and May 9, 1989, 29–30.

48. Krosney, 192.

49. Paul Lewis, "U.N. Chief Says He Will Declare a Cease-Fire in the Iran-Iraq War," *New York Times,* August 2, 1988, sec. A, pp. 1, 9.

50. Yant, 109–111.

51. Said's comment appeared in the *London Review of Books,* March 7, 1991, 7. It was noted by Baghdad-born Kanan Makiya, a human rights activist, who called the Army War College study a "completely discredited document," in *Cruelty and Silence* (New York: Norton, 1993), 348.

52. Stephen C. Pelletiere, *The Iran-Iraq War: Chaos in a Vacuum* (New York: Praeger, 1992), 136–37.

53. Bulloch and Morris, 262; Cordesman and Wagner, 517; Sciolino, 171; Krosney, 123–132; Wright, 174–75.

54. *Bureaucracy of Repression: The Iraqi Government in Its Own Words* (New York: Middle East Watch/Human Rights Watch, February 1994), 10–11.

55. Hiro, 201.

56. Anthony H. Cordesman, *Iran and Iraq: The Threat from the Northern Gulf* (Boulder, CO: Westview), 95.

57. Krosney, 192.

58. William E. Burrows and Robert Windrem, *Critical Mass: The Dangerous Race for Superweapons in a Fragmenting World* (New York: Simon and Schuster, 1994), 49.

59. Flora Lewis, "Move to Stop Iraq," *New York Times,* September 14, 1988, sec. A, p. 31.

60. Robert Pear, "House Panel Seeks to Penalize Iraq for Gas Use," *New York Times,* September 23, 1988, sec. A, p. 9.

61. Quoted in Sciolino, 171.

62. Senate Committee, *Chemical and Biological Weapons Threat,* 30. Unnamed military officials had also been cited in 1988 as concerned that "the extensive use of poison gas in the Iran-Iraq conflict has lowered the threshold of opposition to biological weapons as well." Lee Lescaze, "Quest for Way to Block Biological Weapons Is Itself Called a Threat," *Wall Street Journal,* September 19, 1988, 1.

63. U.S. Senate Committee on Governmental Affairs and the Permanent Subcommittee on Investigations, *Global Spread of Chemical and Biological Weapons: Hearings Before the Committee on Governmental Affairs and the Permanent Subcommittee on Investigations,* 101st Cong., 1st sess., February 9, 10, and May 2, 17, 1989, 7, 16.

64. National Security Directive 26, signed by George Bush on October 2, 1989, reproduced in Friedman, 322.

65. See note 33.

66. *Convention on the Prohibition of the Development, Production and Stockpiling of Bacteriological (Biological) and Toxin Weapons and on Their Destruction,* opened for signature at London, Moscow, and Washington, D.C.: April 10, 1972; entered into force: March 26, 1975.

67. Friedman, 134.

68. Sciolino, 173.

69. Milton Viorst, "Poison Gas and 'Genocide': The Shaky Case Against Iraq," *Washington Post,* October 5, 1988, sec. A., p. 25.

70. Sciolino, 134.

71. Sciolino, 46–47.

72. U.S. House Committee on Foreign Affairs, *Crisis in the Persian Gulf: Hearings Before the Committee on Foreign Affairs,* 101st Cong., 2d sess., September 4, 1990, 17, cited in Freedman and Karsh, 219.

73. Freedman and Karsh, 257.

74. Michael R. Gordon, "C.I.A. Fears Iraq Could Deploy Biological Arms by Early 1991," *New York Times,* September 29, 1990, 4.

75. Malcolm W. Browne, "Army Reported Ready for Iraqi Germ Warfare," *New York Times,* January 6, 1991, 6.

76. Malcolm W. Browne, "Germ Warfare Regarded as a Hard Enemy to Fight," *New York Times,* December 28, 1990, sec. A, p. 6.

77. General H. Norman Schwarzkopf, *It Doesn't Take a Hero* (New York: Bantam, 1993), 509.

78. Respondents could write in additional disruptions under a category for "others," and these received an average rating of 5.0. Only about 200 respondents filled in anything for "others" compared with some 2000 for each of the listed choices. Stanley M. Halpin and S. Delane Keene, *Desert Storm Challenges: An Overview of Desert Storm Survey Responses* (Alexandria, VA: U.S. Army Research Institute for the Behavioral and Social Sciences, January 1993), 11, A-4.

79. The coalition troops' protective gear was superior to that of the Iranians, and the Iranians' beards apparently prevented their masks from working properly. But despite the difference, General Schwarzkopf worried that a gas attack might cause his soldiers to panic.

80. Moore, *A Woman at War,* 172–73.

81. Ibid., 174.

82. Ibid., 105–107.

83. Kathleen C. Bailey, *The UN Inspections in Iraq: Lessons for On-Site Verification* (Boulder, CO: Westview, 1995), 22.

CHAPTER 6

1. The reference to chemical weapons came much earlier in Saddam's speech than did his threat to Israel, and whether he meant a connection between the two is unclear. FBIS-NES-90-064, April 3, 1990, 34–36. See also Gordon M. Burck and Charles C. Flowerree, *International Handbook on Chemical Weapons Proliferation* (New York: Greenwood, 1991), 141–42.

2. "Bonn to Sell Israel 20,000 Gas Masks," *New York Times,* June 2, 1967, 14; "West Germans to Take Back Gas Masks Unused by Israel," *New York Times,* July 1, 1967, 5. Israel's prewar anxiety was heightened by reports that Egypt had used chemicals in the early 1960s against Yemen.

3. Stockholm International Peace Research Institute (SIPRI), *The Problem of Chemical and Biological Warfare,* vol. 2, *CB Weapons Today,* (New York: Humanities Press, 1973), 242.

4. Tony Horwitz, "Israel Prepares for Worst in Chemical War but People Don't Seem Panicked by Threat," *Wall Street Journal,* September 15, 1988, 24.

5. Ariel Levite, "Israel Intensifying Preparations to Counter Chemical Attack," *Armed Forces Journal International*, May 1990, 60.

6. Joel Brinkley, "Israelis' Fear of a Poison Gas Attack Is Growing," *New York Times*, August 24, 1990, sec. A, p. 8; Sabra Chartrand, "Israelis Devise Plastic Suit as Shield Against Iraqi Gas," *New York Times*, August 28, 1990, sec. A, p. 14.

7. Chartrand, op. cit.

8. The Jewish and Arab populations of Israel received masks and defense kits, but the Defense Ministry claimed it did not have enough for the Palestinians in Gaza and the West Bank. The Israeli Supreme Court later ruled that masks and kits be distributed to the Palestinians, although the government was slow to comply. See Jon Immanuel, "UNRWA Asks Several Countries for Gas Masks for Palestinians," *Jerusalem Post* [henceforth *JP*], January 24, 1991, 10; and Moshe Arens, *Broken Covenant* (New York: Simon and Schuster, 1995), 172–73.

9. Liat Collins, "Kits for Protecting House Pets from Gas Attack a 'Big Seller,'" *JP*, January 6, 1991; Bradley Burston, "Gas Masks to Be Issued in Rural Areas," *JP*, January 8, 1991, 1.

10. Bradley Burston, "Q: What If There's War? A: You'll Be All Right," *JP*, January 11, 1991, 7.

11. Bradley Burston, "What to Do If an Alert Is Sounded," *JP*, January 14, 1991, 5.

12. Joel Gordon, "What If a Gas Attack Comes While You're Riding in a Car?" *JP*, January 24, 1991, 2.

13. Burston, "What to Do If an Alert Is Sounded."

14. *Aviation Week*, January 21, 1991, cited by Allison Kaplan, "Defenders Have Two Minutes to Counter Scud," *JP*, January 25, 1991, 1.

15. Bradley Burston, "IDF Medical Chief Plays Down Iraq Chemical, Biological Threat," *JP*, January 14, 1991, 1.

16. Bradley Burston, "Sealed Rooms or Bomb Shelters?" *JP*, February 15, 1991, 2.

17. Bradley Burston, "More Answers to Most-Asked Questions," *JP*, January 16, 1991, 2.

18. Joel Brinkley, "Israel Declares Emergency; Staying Indoors Urged," *New York Times*, January 17, 1991, sec. A, p. 8.

19. "First Scud Attack Victim Buried," *JP*, January 28, 1991, 1.

20. Bradley Burston, "Haga Okays Sealed Shelters," *JP*, February 11, 1991, 1; Burston, "No Injuries as Scud Falls in Empty Area," *JP*, February 12, 1991, 1.

21. Ernie Meyer, "How to Turn a Sealed Room into a Shelter," *JP*, February 17, 1991, 4.

22. David Rudge, "Gas Attack Unlikely for Now, Says Iraq Expert," *JP*, January 21, 1991, 10.

23. Bradley Burston, "Scuds Still Believed to Be Chemical Threat," *JP*, January 25, 1991, 1.

24. Allison Kaplan, "Beware 'Scuds with Gas Warheads'—Cheney," *JP*, January 28, 1991, 1.

25. Bradley Burston and Abraham Rabinovich, "Arens: Iraq Has Crossed the Red Line," *JP*, January 28, 1991, 10.

26. The experience was recounted to me by a chemical and biological defense officer who wished to remain anonymous.

27. Bradley Burston, "Early Scud Caused Gas Scare," *JP*, March 31, 1991, 1.

28. Joel Brinkley, "Iraq Fires New Missile Attack at Israel," *New York Times*, January 19, 1991, 7.

29. Sabra Chartrand, "A Day of False Alarms and Fears, Flanked by Real Explosions," *New York Times*, January 19, 1991, 7.

30. Ibid.

31. Bradley Burston, "Protecting Frightened Children," *JP*, January 20, 1991, 2.

32. Dr. Harats and others commented to me on hospital readiness while I was in Israel from January 30 through February 9, 1991. Also see Y. Shapira et al., "Outline of Hospital Organization for a Chemical Warfare Attack," *Israel Journal of Medical Sciences,* 27, nos. 11–12 (November–December 1991), 616–22.

33. Gil Goldfine, "Paintings from a Sealed Room," *Jerusalem Post Entertainment Magazine,* February 22, 1991, 14.

34. Natalie Saran, "Stern Can't Stop the Music During Alert," *JP*, February 24, 1991, 10. Stern's picture in a gas mask appeared in the *New York Times,* March 3, 1991, sec. H, p. 23, and in *JP*, March 4, 1991, sec. IE, p. 5. The usually unemotional defense minister, Moshe Arens, who was in the audience, later called the episode a "sight I shall never forget." Arens, 210.

35. Judy Siegel-Itzkovich, "From a Sealed Womb to a Sealed Room," *JP*, February 22, 1991, 9.

36. Carl Schrag, "Q. Who Needs Dumb Jokes at a Time Like This? A. We Do," *JP*, February 11, 1991, 5.

37. Martha Meisels, in "War in the Fast Lane," *JP*, February 14, 1991, 5.

38. Ibid., 5.

39. Ruby Karzen, "Solidarity and Solitude," *JP*, March 1, 1991, 10.

40. Shulammit Dovrat, "Spuds 'n' Scuds," *JP*, February 8, 1991, 11.

41. Martha Meisels, "Designer Gas Mask Kits Are 'In'," *JP*, February 4, 1991, 2.

42. General H. Norman Schwarzkopf, *It Doesn't Take a Hero* (New York: Bantam, 1992), 485–86.

43. Arens, 188–89.

44. Schwarzkopf, 486–87.

45. Arens, 215.

46. Ibid., 212–13.

47. Daniella Ashkenazy, "All Quiet on the Home Front," *JP*, March 3, 1991, 5.

48. Gila Svirsky, "Israel Will Be Left with Scars of War," *JP*, February 28, 1991, 4.

49. Peretz Lavie, *The Enchanted World of Sleep* (New Haven, CT: Yale University Press, 1996), 73, 166.

50. The figures are from a preliminary summary in *JP*, March 1, 1991, 2, and subsequent updates, including H. D. Danenberg et al., "Morality in Israel During the Persian Gulf War—Initial Observations," *Israel Journal of Medical Sciences*, 27, nos. 11–12 (November–December 1991), 627–30. Forty-six scuds landed in Saudi Arabia, killing 31 people. Almost all the victims were U.S. personnel, 28 of whom were killed when a missile hit their barracks. Lawrence Freedman and Efraim Karsh, *The Gulf Conflict, 1990–1991: Diplomacy and War in the New World Order* (Princeton, NJ: Princeton University Press, 1993), 307.

51. Clyde Haberman, "Israeli Study Sees Higher Death Rate from '91 Scud Attacks," *New York Times*, April 21, 1995, sec. A, p. 9.

52. Asher Wallfish, "Comptroller Eases Criticism of Army on Gas Masks, but IDF Blasts Her Report," *JP*, April 15, 1991, 7.

53. *JP*, October 15, 1993.

CHAPTER 7

1. Lawrence Freedman and Efraim Karsh, *The Gulf Conflict, 1990–1991* (Princeton, NJ: Princeton University Press, 1993), 416.

2. "Aspin Backs Report of Chemicals in Gulf, but Denies They Caused Vets' Ailments," *Record* (Hackensack, NJ), November 11, 1993, sec. A, p. 23.

3. "France Detected Chemical Agents During Gulf War," *Record* (Hackensack, NJ), December 5, 1993, sec. A, p. 41.

4. Interview on CNN by Peter Arnett, January 28, 1991.

5. *Military Review,* September 1991, 107.

6. Daniel E. Spector, *Military Review,* June 1992, 92.

7. Elaine Sciolino, *The Outlaw State: Saddam Hussein's Quest for Power and the Gulf Crisis* (New York: Wiley, 1991), 262.

8. Elaine Sciolino, "Iraqi Report Says Chemical Arsenal Survived the War," *New York Times,* April 20, 1991, 1.

9. Steven R. Bowman, "Chemical Weapons Proliferation: Issues for Congress," Congressional Research Service, IB90084, Washington, DC, May 5, 1994, 1.

10. United Nations Security Council, *Report of the Secretary-General on the Status of the Implementation of the Special Commission's Plan for the Ongoing Monitoring and Verification of Iraq's Compliance with Relevant Parts of Section C of Security Council Resolution 687 (1991),* (S/1995/284), April 10, 1995, 11–12.

11. "Chemicals Stayed in the Wrappings," *Jerusalem Post,* February 28, 1991, 3.

12. Barbara Crossette, "Iraq Tells U.N. It Understated Germ War Plan," *New York Times,* August 23, 1995, sec. A, p. 1.

13. The passage was in a letter presented by Secretary of State James Baker to Iraq's foreign minister, Tariq Aziz, on January 9, 1991. After reading the text, Aziz refused to accept the letter. The text was conveyed nevertheless through public channels. See Freedman and Karsh, 255.

14. The claim was made by an unnamed "counterterrorism expert," according to K. Scott McMahon, "High-Tech Investments Needed to Counter Smuggled Weapons," *National Defense* 80, no. 514 (January 1996), 28.

15. Freedman and Karsh, 257.

16. Ibid., 331.

17. At the September 1994 Special Conference on the Biological Weapons Convention held in Geneva, I engaged two Iraqi delegates informally. Why had Iraq not used chemical weapons during the Gulf War?

    The men seemed nonplussed, as if this were the least expected question anyone might ask. After a pause, Khalid Al-Khero softly responded, "We had to play by the rules." The other Iraqi, Mohammed Hussein, apparently superior in rank, interrupted: "We had to consider the reflection from the other side."

You mean the other side might react angrily? "Yes, exactly."

But Iraq possessed chemical weapons; could they not have been used effectively? Al-Khero: "We had them but it was not wise to use them because they would not have been effective."

They were effective during the war with Iran, weren't they?

"We did not use them against Iran," Hussein and Al-Khero replied almost in unison.

There were press accounts, reports to the United Nations, I said. I reminded them of the Iraqi foreign minister's admission to reporters at the end of the Iran-Iraq war that they had been used.

Both: "No, no. There is no proof they were used." The men made clear that they were not interested in further discussion.

18. Donald Mahley, interview, Geneva, September 20, 1994.

19. Graham Pearson, interview, Geneva, September 21, 1994.

20. James A. Baker III, *The Politics of Diplomacy: Revolution, War and Peace, 1989–1992* (New York: Putnam, 1995), 359.

21. Freedman and Karsh, 434.

22. R. Jeffrey Smith, "Iraq Admits It Prepared Fatal Toxins for Gulf War," *Record* (Hackensack, NJ), August 26, 1995, sec. A, p. 12.

23. U.S. Senate Committee on Veterans' Affairs, *Persian Gulf War Illnesses: Are We Treating the Veterans Right? Hearing Before the Committee on Veterans' Affairs,* 103d Cong., 1st sess., November 16, 1993, 46–49.

24. National Institute of Medicine, *Health Consequences of Service During the Persian Gulf War: Initial Findings and Recommendations for Immediate Action* (Washington, DC: National Academy Press, 1995), 36.

25. Dave Parks, "Battling an Unseen Foe," and "Navy Man Who May Have Parasite Says Life Ruined," *Birmingham News,* October 18, 1992. These articles appeared on pages 6 and 7 of a 1994 supplement of the newspaper titled *Desert Storm Diseases.* The supplement contained some 75 stories on the subject that had appeared in the newspaper between October 1992 and December 1993.

26. Senate Committee, *Persian Gulf War Illnesses,* 29–32. Smith elaborated on his condition on the *Donahue* show on "Germ Warfare in Desert Storm," NBC-TV, air date: March 23, 1994.

27. Senate Committee, *Persian Gulf War Illnesses,* 49–50.

28. Dave Parks, "Ill Vets Say Wives Show Similar Symptoms," *Birmingham News,* January 28, 1993, suppl. 9; "Vets' Babies Suffer Health

Woes," *Birmingham News,* November 20, 1993, suppl. 29; Senate Committee, *Persian Gulf War Illnesses,* 48.

29. Dave Parks, "Fallout Fears," *Birmingham News,* April 25, 1993, suppl. 12; Dave Parks, "Senators Probing Veterans' Chemical Attack Claims," *Birmingham News,* June 30, 1993, suppl. 18.

30. Dave Parks, "Living with Scars," *Birmingham News,* December 12, 1993, suppl. 3.

31. Michael Brumas, "Gulf Veterans' Ills Called 'Agent Orange Revisited,'" *Birmingham News,* June 9, 1993, suppl. 15.

32. The Vietnam veterans' health complaints began in 1978, three years after the American withdrawal from the area. Agent Orange is composed of dioxin and other toxic ingredients. Nevertheless, the Department of Defense insisted that the material did not cause serious health problems to people. Claims for service-connected disability were routinely denied. In his book on the Agent Orange issue, Fred Wilcox noted that "veterans seeking help from VA hospitals were often given inadequate, sometimes hostile, treatment and diagnosed as suffering from a variety of psychosomatic symptoms." Fred A. Wilcox, *Waiting for an Army to Die: The Tragedy of Agent Orange* (New York: Vintage, 1983), xii.

33. "The Agent Orange Case," *New York Times,* April 4, 1995, sec. D, p. 7.

34. Office of the Secretary of Defense, William J. Perry and John M. Shalikashvili, Memorandum for Persian Gulf War Veterans, Subject: Persian Gulf War Health Issues, May 25, 1994.

35. Veterans' Benefits Improvement Act of 1994, Title 1: Persian Gulf War Veterans' Benefits Act, Public Law 103-446, 108 Stat. 4645.

36. "Vets Sue Companies for Selling Precursors to Chemical Arms," *Chemical and Engineering News* 72, no. 25 (June 25, 1994), 21.

37. U.S. Senate, *U.S. Chemical and Biological Warfare-Related Dual Use Exports to Iraq and Their Possible Impact on the Health Consequences of the Persian Gulf War,* committee staff report (no. 3) to Donald W. Riegle, Jr., chairman of the Committee on Banking, Housing and Urban Affairs with respect to export administration, October 7, 1994, 1–2.

38. U.S. Senate, *U.S. Chemical and Biological Warfare-Related Dual Use Exports to Iraq and Their Possible Impact on the Health Consequences of the Persian Gulf War,* a report by Donald W. Riegle, Jr., chairman, and Alfonse M. D'Amato, ranking member, of the Committee on Banking, Housing and Urban Affairs with respect to export administration, May 25, 1994, 4.

39. Ibid., 150.

40. Ibid., 57–58.

41. Ibid., 150. In 1966, the Pentagon said "new information" indicated that perhaps 400 American troops may have been exposed to nerve gas when an Iraqi chemical munitions depot was destroyed. Philip Shenon, "Gulf War Illness May Be Linked to Gas Exposure, Pentagon Says," *New York Times,* June 22, 1996, 1.

42. Senate Committee staff report, October 7, 1994, 19–20.

43. Daniel Williams, "Senator: US Is Hiding Evidence of Chemicals Used in Gulf War," *Jerusalem Post,* October 11, 1994, 4, reprinted from the *Washington Post.*

44. U.S. Senate, *Is Military Research Hazardous to Veterans' Health? Lessons Spanning Half a Century,* a staff report prepared for the Committee on Veterans' Affairs, December 8, 1994, 26.

45. U.S. Senate Committee on Veterans' Affairs, *Is Military Research Hazardous to Veterans' Health? Lessons from World War II, the Persian Gulf, and Today: Hearing Before the Committee on Veterans' Affairs,* 103d Cong., 2d sess., May 6, 1994, 8–10.

46. Ibid., 9.

47. Ibid., 431–32. Pentagon officials later estimated that perhaps 250,000 troops took some pyridostigmine bromide but conceded that they did not know the amounts taken or the number of troops who took any at all. *Presidential Advisory Committee on Gulf Veterans' Illnesses: Interim Report* (Washington, DC: Government Printing Office, February 1996), 21–22.

48. Senate Committee, *Is Military Research Hazardous to Veterans' Health?* 34–35.

49. Ibid., 41.

50. Ibid., 48.

51. Senate Committee, *Is Military Research Hazardous to Veterans' Health?* staff report, 11.

52. Ibid., 29–31. The unpublished study, titled "The Effects of Pyridostigmine Bromide on Vision," was provided to the army's Human Use Review Board on August 15, 1990, and is in the files of the Senate Committee on Veterans' Affairs.

53. *Physicians' Desk Reference,* 48th ed. (Montvale, NJ: Medical Economics Data Production Company, 1994), 1051.

54. Senate Committee, *Is Military Research Hazardous to Veterans' Health?* 30.

55. National Institute of Medicine, 54–55.

56. Ibid., 56–59.

57. John F. Harris, "Clinton Woos Lukewarm Veterans," *Washington Post,* March 7, 1995, sec. A, p. 7.

58. The findings were based on exposure of hens to the chemicals. Mohamed B. Abou-Donia et al., "Neurotoxicity Results from Coexposure to Pyridostigmine Bromide, Deet, and Permethrin," *Journal of Toxicology and Environmental Health* 48, no. 1 (May 1996), 1–22.

59. U.S. Department of Defense, Comprehensive Clinical Evaluation Program for Gulf War Veterans, Washington, DC, August 1995, 16, 45–48.

60. Todd S. Purdom, "Hillary Clinton Urges Attention to Gulf War Ailments," *New York Times,* August 15, 1995, sec. A, p. 15. *Presidential Advisory Committee on Gulf Veterans' Illnesses,* 2.

61. Kenneth Miller and Jimmie Briggs, "The Tiny Victims of Desert Storm," *Life,* November 1995, 52.

62. Diana Zuckerman, interview, May 1, 1995. In later discussion, on June 15, 1995, Zuckerman said she and another committee staff member, Patricia Olson, had reviewed theories about possible causes of Gulf War syndrome. They were surprised to see that, in earlier studies of pyridostigmine with healthy men, several had severe reactions. Senator Rockefeller chaired hearings on the subject in 1994. Resulting legislation requires the Defense Department to study the effects of the drug.

63. Ian Johnstone, *Aftermath of the Gulf War: An Assessment of UN Action* (Boulder, CO: Lynne Rienner, 1994), 15.

64. UN Security Council Resolution 687, April 3, 1991.

65. "Excerpts from Letter to U.N.," *New York Times,* April 8, 1991, sec. A, p. 6.

66. Johnstone, 29.

67. Statement by the president of the UN Security Council (S/22746), June 28, 1991.

68. Resolution 707, August 15, 1991; statements by the president of the UN Security Council to the press, September 23 and 24, 1991; statement by the president of the UN Security Council (S/23663), February 28, 1992.

69. Statement by the president of the UN Security Council (S/24836), November 23, 1992.

70. Johnstone, 30.

71. Kathleen C. Bailey, *The UN Inspections in Iraq: Lessons for On-Site Verification* (Boulder, CO: Westview, 1995), 1.

72. Ibid., 1.

73. Statements by the president of the UN Security Council (S/23699), March 11, 1992, and (S/24836), November 23, 1992.

74. Bailey, 11–12.

75. Ibid., 13.

76. Ibid., 39–40.

77. Ibid., 38–39. John R. Walker, "UN Resolution 687: Can Iraq's BW Programme Be Eliminated?" paper presented at Conference on Controlling Biological Weapons, Wilton Park, United Kingdom, September 24–26, 1993.

78. UN Security Council, *Report of the Secretary-General* (S/1995/284), April 10, 1995, 5.

79. Ibid., 16.

80. Ibid., 17.

81. Ibid.

82. Ibid., 17–18.

83. Ibid., 19.

84. Ibid., 21.

85. Ibid., 22.

86. Edited versions appear in Raymond A. Zilinskas, ed., "Symposium of United Nations Biological Weapons Inspectors: Implications of the Iraqi Experience for Biological Arms Control," *Politics and the Life Sciences* 14, no. 2 (August 1995), 229–62.

87. Anna Johnson-Winegar, "The Role of Declarations in UNSCOM's Program in Iraq," *Politics and the Life Sciences* 14, no. 2 (August 1995), 238.

88. David L. Huxsoll, "On-site Inspection Measures and Interviews," *Politics and the Life Sciences* 14, no. 2 (August 1995), 239.

89. Raymond A. Zilinskas, "UNSCOM and the UNSCOM Experience in Iraq," *Politics and the Life Sciences* 14, no. 2 (August 1995), 232.

90. UN Security Council, *Report of the Secretary-General*, 5.

91. William Safire, "Iraq's Ton of Germs," *New York Times,* April 13, 1995, 25.

92. Christopher S. Wren, "U.N. Expert Raises Estimates of Iraq's Biological Arsenal," *New York Times,* June 21, 1995, sec. A, p. 6.

93. R. Jeffrey Smith, "Iraq Had Program for Germ Warfare," *Washington Post,* July 6, 1995, sec. A, pp. 1, 17.

94. Barbara Crossette, "Iraq Tells U.N. It Understated Germ War Plan," *New York Times,* August 23, 1995, sec. A, p. 1. Despite Iraq's handing over to UNSCOM thousands of previously concealed documents at the end of 1995, the disclosures remained incomplete. UN Security Council, *Tenth Report of the Executive Chairman of the Special Commission* (S/1995/1038), December 17, 1995. Ostensibly forgiven by their father-in-law, Hussein Kamel Hassan and his brother returned with their wives to Iraq on February 20, 1996. Two days later the wives divorced them. The next day the brothers were murdered, reportedly by relatives. Douglas Jehl, "Iraqi Defectors Killed 3 Days After Return," *New York Times,* February 24, 1996, sec. A, p. 1.

CHAPTER 8

1. Jonathan Annells and James Adams, "Did Terrorists Kill with a Deadly Nerve Gas Test?" *Sunday Times* (London), March 19, 1995; *Mainichi Daily News* (Tokyo), July 9, 1994.

2. CNN News, June 28, 1994; *Mainichi Daily News,* July 16, 1994.

3. *Mainichi Daily News,* June 30, 1994.

4. *Mainichi Daily News,* July 2, 1994.

5. Keiichi Tsuneishi is quoted in the *Mainichi Daily News,* July 4, 1994.

6. *Mainichi Daily News,* July 9, 1994.

7. *Sydney Morning Herald,* August 20, 1994.

8. Kyle B. Olson, "The Matsumoto Incident: Sarin Poisoning in a Japanese Residential Community," Chemical and Biological Arms Control Institute, Alexandria, VA, December 1994, 5.

9. Ibid., 4.

10. Ibid., 5.

11. Annells and Adams.

12. Lisa Rein, Joe Nicholson, and Al Baker, "New York's Not Prepared," *Daily News* (New York), March 21, 1995, 1, 7.

13. "Passengers Released from Plane Held at Kennedy," Associated Press Wire Service, March 21, 1995.

14. Thomas J. Fitzgerald, *Record* (Hackensack, NJ), March 21, 1995, sec. A, p. 10.

15. Jane H. Lii, "In Subways Elsewhere, a Heightened State of Alert," *New York Times,* March 21, 1995, sec. A, p. 12.

16. Nicholas D. Kristof, "Japanese Police Find Chemicals and Gas Masks at Sect's Offices," *New York Times,* March 23, 1995, sec. A, p. 1; Kristof, "Japanese Indict Leader of Cult in Gas Murders," *New York Times,* June 7, 1995, sec. A, p. 6.

17. U.S. Senate Permanent Subcommittee on Investigations, *Global Proliferation of Weapons of Mass Destruction: Hearings Before the Permanent Subcommittee on Investigations,* 104th Cong., 1st sess., Minority Staff Statement, "A Case Study on the Aum Shinrikyo," Washington, DC, October 31, 1995, 52.

18. Andrew Pollack, "Japanese Police Say They Found Germ-War Material in Cult Site," *New York Times,* March 29, 1995, sec. A, p. 10.

19. Senate Permanent Subcommittee, "A Case Study on the Aum Shinrikyo," 43.

20. Kristof, "Japanese Police Find Chemicals."

21. Andrew Pollack, "The Scent of Terror," *New York Times,* April 23, 1995, sec. E, p. 2.

22. Nicholas D. Kristof, "Japanese Police Foil Gas Attack in Subway," *New York Times,* May 7, 1995, 10; Kristof, "How Tokyo Barely Escaped Even Deadlier Subway Attack," *New York Times,* May 18, 1995, sec. A, p. 14; Senate Permanent Subcommittee, "A Case Study on Aum Shinrikyo," 56.

23. Michael Janofsky, "Looking for Motives in Plague Case," *New York Times,* May 28, 1995, 18.

24. Ibid.

25. Ibid.

26. Initially scheduled for September in Dayton, Ohio, the federal court trial date was changed to November. "September Trial Is Slated in Plague Case," *Columbus Dispatch* (Ohio), July 11, 1995; "A Guilty Plea in Mail-Order Bacteria Case," *New York Times,* November 24, 1995, sec. A, p. 23. In another case, two members of the Minnesota Patriots Council, a militia group, were convicted in March 1995 of planning to attack government officials with the toxin ricin. Kenneth S. Stern, *A Force upon the Plain: The American Militia Movement and the Politics of Hate* (New York: Simon and Schuster, 1996), 135.

27. An antiterrorism law was enacted in April 1996 that included provisions to regulate transfers of biological agents that "pose a severe threat to public health and safety" (PL 104-132), 74–75. How effective the provisions will be remains uncertain.

28. U.S. Senate Committee on Governmental Affairs, *Global Spread of Chemical and Biological Weapons: Hearings Before the Committee on Governmental Affairs and Its Permanent Subcommittee on Investigations,* 101st Cong., 1st sess., February 9, 10 and May 2, 17, 1989, 2.

29. J. P. Perry Robinson, "The Chemical Industry and Chemical-Warfare Disarmament: Categorizing Chemicals for the Purposes of the Projected Chemical Weapons Convention," in *The Chemical Industry and the Projected Chemical Weapons Convention,* Proceedings of SIPRI/Pugwash Conference (New York: Oxford University Press, 1986); W. Seth Carus, *The Genie Unleashed: Iraq's Chemical and Biological Weapons Program* (Washington, DC: The Washington Institute for Near East Policy, 1989), chaps. 2 and 3; Herbert Krosney, *Deadly Business: Legal Deals and Outlaw Weapons, the Arming of Iran and Iraq, 1975 to the Present* (New York: Four Walls Eight Windows, 1993), 55–56.

30. Edward Naidus, interview, June 1, 1995. Much of this section is drawn from Leonard A. Cole, "At Home with Biological and Chemical Weapons," *Medicine and Global Survival* 2, no. 3 (September 1995), 182–83.

31. David Goldberg, *Global Spread of Chemical and Biological Weapons: Hearings,* 34–35.

32. John Cookson and Judith Nottingham, *A Survey of Chemical and Biological Warfare* (New York: Monthly Review Press, 1969), 276.

33. Nancy Connell, interview, May 31, 1995.

34. In such a scenario, if anthrax spores were released at a rate of 2 kilograms per hour, more than 400,000 people would die within 48 hours, according to Robert H. Kupperman and David M. Smith, "Coping with Biological Terrorism," in *Biological Weapons: Weapons of the Future?* ed. Brad Roberts (Washington, DC: Center for International and Strategic Studies, 1993), 42. A similar doomsday script is in Neil C. Livingstone, "Biological Nightmares," *Sea Power,* April 1995, 90.

35. Kyle Olson, interview, June 8, 1995.

36. Kupperman and Smith, 38.

37. U.S. Office of Technology Assessment, *Technology Against Terrorism: Structuring Security* (Washington, DC: Government Printing Office, January 1992), 40.

38. U.S. Office of Technology Assessment, *Technology Against Terrorism: The Federal Effort* (Washington, DC: Government Printing Office, July 1991), 16.

39. Ibid., 16–17.

40. Richard E. Rubenstein, *Alchemists of Revolution: Terrorism in the Modern World* (New York: Basic Books, 1987), 228. In his chapter 4, Rubenstein disputes such network theorists as Claire Sterling, *The Terror Network: The Secret War of International Terrorism* (New York: Holt, Rinehart and Winston, 1981), and Ray S. Cline and Yonah Alexander, *Terrorism: The Soviet Connection* (New York: Crane Russak, 1984), and such permissive-society theorists as Walter Laqueur, *Terrorism* (Boston: Little, Brown, 1977), and Paul Wilkinson, *Terrorism and the Liberal State* (New York: Wiley, 1977).

41. Rubenstein, 6.

42. David E. Long, *The Anatomy of Terrorism* (New York: Free Press, 1990), 10.

43. Brian M. Jenkins, "The Terrorist Mindset and Terrorist Decision-making" (Santa Monica, CA: Rand, June 1979), 5, cited in Long, 10.

44. Long, 131.

45. Ibid.

46. Walter Laqueur, *The Age of Terrorism* (Boston: Little, Brown, 1987), 317–18.

47. Neil C. Livingstone and Joseph D. Douglass, Jr., *CBW: The Poor Man's Atomic Bomb* (Cambridge, MA: Institute for Foreign Policy Analysis, February 1984), 22.

48. U.S. Office of Technology Assessment, *Technology Against Terrorism: Structuring Security,* 39–40.

49. Long, 131–32.

50. Ibid., 132.

51. Brian M. Jenkins, "Will Terrorists Go Nuclear," in *The Terrorism Reader,* rev. ed., ed. Walter Laqueur and Yonah Alexander (New York: NAL Penguin, 1987), 353. Moral qualms are rarely cited as a reason for the infrequency of chemical and biological terrorism, as confirmed in a recent literature review. Analysts give more credence to such presumptions as lack of control of such weaponry or concerns by terrorists for their own safety. The unpublished review, titled "Chemical and Biological Terrorism: The Threat According to the Open Literature," was prepared in June 1995 by Ron Purver, strategic analyst for the Canadian Security Intelligence Service. The overall subject is replete with conjecture and short on convincing evidence.

52. Ibid., 355.

53. Brian Jenkins, presentation at ChemBio Terrorism Conference, sponsored by the Chemical and Biological Arms Control Institute,

Washington, DC, April 29, 1996. Also see statement of Milton Leitenberg, Senate Permanent Subcommittee, *Global Proliferation of Weapons of Mass Destruction: Hearings Before the Permanent Subcommittee on Investigations,* November 1, 1995, 10.

54. Al O'Leary, *Dateline NBC,* March 21, 1995.

55. Lawrence K. Altman, "Plan Drawn to Help Fight Poison Attack," *New York Times,* March 26, 1995, 9.

56. Other support bodies also are listed in the nation's medical emergency response plan, including the American Red Cross, the Environmental Protection Agency, and the Departments of Agriculture, Justice, and Transportation. *The Federal Response Plan,* for Public Law 93-288, as amended, April 1992, ESF no. 8.

57. *The Federal Response Plan,* ESF no. 8-12. The response action for chemical hazards is virtually the same.

58. James Rabb, interview, June 7, 1995.

59. Kupperman and Smith, 35.

60. Ibid., 43–44.

61. Robert Kupperman, interview, June 17, 1995.

62. "A Study of the Vulnerability of Subway Passengers in New York City to Covert Action with Biological Agents," Miscellaneous Publication 24, Department of the Army, Fort Detrick, Frederick, Maryland, January 1968, 7.

63. Ibid., 26–27.

64. Senate Permanent Subcommittee, *Global Spread of Chemical and Biological Weapons: Hearings,* 203.

CHAPTER 9

1. Rosenberg's remarks are from interviews on September 20, 1994, and September 5, 1995.

2. Mahley's remarks are from interviews on September 20, 1994, and July 24, 1995.

3. Kathleen Bailey, interviews, July 13 and 31, 1995. Also see Kathleen C. Bailey, "Why the Chemical Weapons Convention Should Not Be Ratified," in *Ratifying the Chemical Weapons Convention,* ed. Brad Roberts (Washington, DC: Center for Strategic and International Studies, 1994), 58.

4. Kathleen C. Bailey, *The UN Inspections in Iraq: Lessons for On-Site Verification* (Boulder, CO: Westview, 1995); K. C. Bailey, *Death for Cause* (Livermore, CA: Meerkat Publications, 1995).

5. U.S. Senate Committee on Foreign Relations, *Chemical Weapons Convention (Treaty Doc. 103-21): Hearings Before the Committee on Foreign Relations,* 103d Cong., 2d sess., March 22, April 13, May 13 and 17, June 9 and 23, 1994, 123. Frank Gaffney, interview, August 9, 1995.

6. *Convention on the Prohibition of the Development, Production and Stockpiling of Bacteriological (Biological) and Toxin Weapons and on Their Destruction,* opened for signature at London, Moscow, and Washington, D.C.: April 10, 1972; entered into force: March 26, 1975; henceforth referred to as the Biological Weapons Convention (BWC).

7. "Biological Weapons: 70 Nations Say No," *New York Times,* April 16, 1972, sec. 4, p. 6.

8. Remarks of the president on announcing the chemical and biological defense policies and programs, Office of the White House Press Secretary, November 24, 1969.

9. The White House, news release, February 14, 1970.

10. David Z. Beckler, interview, June 9, 1995. Marie Isabelle Chevrier, "Verifying the Unverifiable: Lessons from the Biological Weapons Convention," *Politics and the Life Sciences* 9, 1 (August 1990), 97.

11. Hans Swyter, quoted in Susan Wright, ed., *Preventing a Biological Arms Race* (Cambridge, MA: MIT Press, 1990), 40.

12. Robert Harris and Jeremy Paxman, *A Higher Form of Killing: The Secret Story of Chemical and Biological Warfare* (New York: Hill and Wang, 1992), 171–72.

13. Wright, 41; Hugh D. Crone, *Banning Chemical Weapons: The Scientific Background* (New York: Cambridge University Press, 1992), 108.

14. Nicholas A. Sims, *The Diplomacy of Biological Disarmament: Vicissitudes of a Treaty in Force, 1975–85* (New York: St. Martin's, 1988), 255.

15. "Biological Weapons: Debate on Anthrax Incident Flares," *Chemical and Engineering News* 65, no. 4 (April 6, 1987), 4; John H. Cushman, Jr., "Russians Explain '79 Anthrax Cases," *New York Times,* April 14, 1988, sec. A, p. 7.

16. Matthew Meselson et al., "The Sverdlovsk Anthrax Outbreak of 1979," *Science* 266, no. 5188 (November 18, 1994), 1207. Before Meselson investigated the Sverdlovsk site in 1992, he had been skeptical about U.S. claims of a military connection to the out-

break. Finding the earlier Soviet denials plausible, he called for "objective review" of the U.S. version. See his article titled "The Biological Weapons Convention and the Sverdlovsk Anthrax Outbreak of 1979," *F.A.S.* (*Journal of the Federation of American Scientists*) 41, no. 7 (September 1988), 1–4.

17. Lois Ember, "Yellow Rain," *Chemical and Engineering News* 62, no. 2 (January 9, 1984), 10–18.

18. Thomas D. Seeley, Joan W. Nowicke, Matthew Meselson, Jeanne Guillemin, and Pongthep Akratanakul, "Yellow Rain," *Scientific American* 253, no. 3 (September 1985), 137. See also Julian Robinson, Jeanne Guillemin, and Matthew Meselson, "Yellow Rain: The Story Collapses," *Foreign Policy* 68 (Fall 1987); Elisa Harris, "Sverdlovsk and Yellow Rain: Two Cases of Noncompliance?" *International Security* 11, no. 4 (Spring 1987).

19. Sims, 174–76.

20. Barend ter Haar, *The Future of Biological Weapons* (New York: Praeger, 1991), 28–30.

21. Ibid., 34–35.

22. Nicholas Sims, "Achievements and Failures at the Third Review Conference," *Chemical Weapons Convention Bulletin,* no. 14 (December 1991), 2–3.

23. CD/294, July 21, 1982, sec. III.

24. Valerie Adams, *Chemical Warfare, Chemical Disarmament* (Bloomington: Indiana University Press, 1990), 178–79.

25. Ibid., 179–80.

26. CD/500, April 18, 1984, article XI.

27. Adams, 180.

28. United Kingdom, Foreign and Commonwealth Office, *Chemical Weapons Convention Negotiations, 1972–92,* foreign policy document no. 243, July 1993, 5.

29. Adams, 183.

30. Adams, 184–85.

31. Ibid., 192.

32. Michael E. Gordon, "Moscow to Begin Destroying Its Chemical Weapons," *New York Times,* January 9, 1989, sec. A, p. 8.

33. James M. Markham, "Arabs Link Curbs on Gas and A-Arms," *New York Times,* January 9, 1989, sec. A, p. 8.

34. Some think the participants were justified in their reluctance to criticize Iraq from the podium. If they had, the Arab League might

not have joined in the final declaration that reaffirmed the Geneva Protocol and called for accelerating the Chemical Weapons Convention negotiations. Julian Perry Robinson, personal communication, January 26, 1996.

35. *New York Times,* September 26, 1989, sec A., p. 16.

36. "The Two-Percent Solution," *Chemical Weapons Convention Bulletin,* no. 7 (February 1990), 1.

37. *Agreed Statement in Connection with the Agreement Between the United States of America and the Union of Soviet Socialist Republics on Destruction and Non-Production of Chemical Weapons and on Measures to Facilitate the Multilateral Convention on Banning Chemical Weapons.* Reproduced in *Chemical Weapons Convention Bulletin,* no. 8 (June 1990), 22.

38. H. Martin Lancaster, "Chemical Proliferation and Disarmament: A Congressional Perspective," *Chemical Weapons Convention Bulletin,* no. 11 (March 1991), 1.

39. Ibid., 3.

40. Ibid., 4

41. *Chemical Weapons Convention Negotiations, 1972–1992,* 21.

42. *Convention on the Prohibition of the Development, Production, Stockpiling, and Use of Chemical Weapons and on Their Destruction,* opened for signature at Paris, January 13, 1993. Processes and forces behind the development of the CWC are examined in Thomas Bernauer, *The Chemistry of Regime Formation,* United Nations Institute for Disarmament Research (Brookfield, VT: Dartmouth, 1993).

43. Barbara Crossette, "Chemical Treaty Appears on Hold in the Senate," *New York Times,* September 10, 1995, 12. Letter to President Bill Clinton from Senator Jesse Helms, October 25, 1995.

44. Chevrier, 93–105.

45. Report, Ad Hoc Group of Governmental Experts to Identify and Examine Potential Verification Measures from a Scientific and Technical Standpoint, BWC/Conf. II/VEREX/9, Geneva, 1993.

46. Ibid., 11–20.

47. Ibid., 9.

48. Proposal for a Mandate for an Ad-Hoc Working Group on Verification, BWC/SPCONF/WP.1, September 20, 1994.

49. Statement by Richard Starr at the Biological Weapons Convention Special Conference, Geneva, September 19, 1994.

50. Statement by Soemadi D. M. Brotodiningrat at the Biological Weapons Convention Special Conference, Geneva, September 20, 1994.

51. Statement by Donald Mahley at the Biological Weapons Convention Special Conference, Geneva, September 19, 1994.

52. Final Declaration, BWC/SPCONF/DC/WP.2, September 28, 1994. For an overview of the proceedings, see Marie Isabelle Chevrier, "From Verification to Strengthening Compliance: Prospects and Challenges of the Biological Weapons Convention," *Politics and the Life Sciences* 14, no. 2 (August 1995).

53. In offering the list, Roque Monteleone Neto acknowledged that additional microorganisms could have been included. "Criteria for the Establishment of the First List of Agents," paper presented at Beyond VEREX, a forum sponsored by the Federation of American Scientists and the Special NGO [Non-Governmental Organization] Committee for Disarmament, Geneva, September 21, 1994.

54. "Estimate of the International Costs for a BWC Compliance Regime," FAS Working Group on Biological Weapons Verification, August 1994, paper presented by Beyond VEREX, September 21, 1994; revised October 1994.

55. Robert Mikulak, "What Needs to Be Verified?" *Implementing a Global Chemical Weapons Convention,* proceedings of a 1989 annual meeting symposium, American Association for the Advancement of Science (Washington, DC: AAAS, 1989), 5, 31–32.

56. Senate Committee, *Chemical Weapons Convention,* 107.

57. Sidney N. Grabeal and Patricia Bliss McFate, "The Role of Verification in Arms Control," in *Technology, Security, and Arms Control for the 1990s,* ed. Elizabeth Kirk (Washington, DC: AAAS, 1988), 231–32.

58. Vil S. Mirzayanov, "Dismantling the Soviet/Russian Chemical Weapons Complex: An Insider's View," in *Chemical Weapons Disarmament in Russia: Problems and Prospects,* by Amy E. Smithson, Dr. Vil S. Mirzayanov, Maj. Gen. Roland Lajoie (USA, Ret.), and Michael Krepon, rep. no. 17 (Washington, DC: Henry L. Stimson Center, October 1995), 31.

59. Milton Leitenberg, "Biological Weapons Arms Control," *Contemporary Security Policy* (London) 17, no. 1 (April 1966).

60. Andrew Lawler, "Libya and Iran Seek Ex-Soviet Scientists," *Science* 271, no. 5255 (March 15, 1996), 1485.

61. A report for the UN secretary general in 1969 estimated that a chemical-biological civil defense program could exceed $20 billion.

(Subsequent inflation means the figure would now be about $80 billion.) Stockholm International Peace Research Institute (SIPRI), *The Problem of Chemical and Biological Warfare: C.B. Weapons Today*, vol. 2 (New York: Humanities Press, 1973), 106. Jonathan Tucker presents as thoughtful a case as can be made for a civilian defense program against chemical and biological terrorism: "Chemical/Biological Terrorism: Coping with a New Threat," *Politics and the Life Sciences* 15, 2 (September 1996). Some of his suggestions, including enhancement of intelligence and surveillance, are unarguable. Others, such as stockpiling defense materials at regional facilities seemed especially urgent after anthrax bioterrorism in 2001. Although nerve agents can kill in minutes and a biological attack might not be evident before people are beyond cure, a defense program could help many. The anthrax attacks showed that antibiotic treatment could be effective even after symptoms appear. True, even belated medical care might help some people, but the benefits of a defense program against an unexpected attack are speculative.

62. Senate Committee, *Chemical Weapons Convention*, 134.

CHAPTER 10

1. Graham S. Pearson, "Verification of the Biological Weapons Convention," in *The Verification of the Biological Weapons Convention: Problems and Perspectives,* ed. Oliver Thränert (Bonn: Gustav-Stresemann-Institut, 1992), 92.

2. Graham S. Pearson, "Prospects for Chemical and Biological Arms Control: The Web of Deterrence," *The Washington Quarterly* 16, no. 2 (Spring 1993).

3. Graham S. Pearson, "Strengthening the Biological and Toxin Weapons Convention: The Need for a Verification Protocol," *Industry Insights* (published by the Chemical and Biological Arms Control Institute), no. 3 (September 1994), 4. He includes the phrase with similar descriptions in "Biological Weapons: The British View," in *Biological Weapons: Weapons of the Future?* ed. Brad Roberts (Washington, DC: Center for Strategic and International Studies, 1993), 13–14; "The BWC Special Conference: A British Perspective," paper presented at the Chemical and Biological Arms Control Institute, Washington, DC, June 24, 1994; "Strengthening the Biological Weapons Convention: National and International Security," paper presented at Beyond VEREX, a forum sponsored by the Federation of American Scientists and the Special NGO [Non-Governmental Organization] Committee for Disarmament,

Geneva, September 21, 1994; and "Vaccines for Biological Defence: Defence Considerations," in *Control of Dual-Threat Agents: The Vaccines for Peace Program,* ed. Erhard Geissler and John P. Woodall (New York: Oxford University Press, 1994), 154.

4. Federation of American Scientists, Proposals for Technological Cooperation to Implement Article X of the Biological Weapons Convention, working paper, July 1995, 2–3.

5. Pearson, "Biological Weapons: The British View," 15.

6. U.S. Office of the Secretary of Defense, "Report of the Secretary of Defense's Ad Hoc Committee on Biological Warfare," July 11, 1949, 12.

7. Richard M. Clendenin, *Science and Technology at Fort Detrick, 1943–1968* (Frederick, MD: Fort Detrick, Historian, Technical Information Division, April 1968), x–xi.

8. David L. Huxsoll, Cheryl D. Parrott, and William C. Patrick III, "Medicine in Defense Against Biological Warfare," *JAMA (Journal of the American Medical Association)* 262, no. 5 (August 4, 1989), 679.

9. Thomas R. Dashiell, "The Need for a Defensive Biological Research Program," *Politics and the Life Sciences* 9, no. 1 (August 1990), 88.

10. Harlee Strauss and Jonathan King, "The Fallacy of Defensive Biological Weapon Programmes," in *Biological and Toxin Weapons Today,* ed. Erhard Geissler (New York: Oxford University Press, 1986), 67.

11. Ibid., 68.

12. Jonathan B. Tucker, "Dilemmas of a Dual-Use Technology: Toxins in Medicine and Warfare," *Politics and the Life Sciences* 13, no. 1 (February 1994), 56.

13. Susan Wright and Stewart Ketcham, "The Problem of Interpreting the U.S. Biological Defense Research Program," in *Preventing a Biological Arms Race,* ed. Susan Wright (Cambridge, MA: MIT Press, 1990), 180.

14. Huxsoll et al., 677–79.

15. Michael E. Frisina, "The Offensive-Defensive Distinction in Military Biological Research," *Hastings Center Report* 20, no. 3 (May/June 1990), 21.

16. Ibid., 20–21.

17. Col. John Doesburg, manager of the Joint Program Office for Biological Defense, cited in Christopher Parent, "DOE Eliminates Vac-

cine Production Funding from Biological Defense Office," *Inside the Pentagon* (Arlington, VA), February 1, 1996, 2.

18. Seth Shulman, *Biohazard: How the Pentagon's Biological Warfare Research Program Defeats Its Own Goals* (Washington, DC: Center for Public Integrity, 1993), 2. See also Meryl Nass, "The Labyrinth of Biological Defense," *The PSR Quarterly* 1, no. 1 (March 1991), 25.

19. Raymond Zilinskas, ed., *The Microbiologist and Biological Defense Research: Ethics, Politics, and International Security* (New York: New York Academy of Sciences, 1992), xii.

20. Ibid., 95, 103, 116–117, 133.

21. Ibid., 43.

22. Nerve agents such as sarin or VX can penetrate ordinary clothing, and a protective suit must be worn along with a gas mask containing an activated charcoal filter. The relatively large particle size of biological agents allows for protection with less cumbersome respirator masks than required for gas. Because biological agents cannot usually penetrate intact skin, special outerwear is less essential. (Nevertheless, organisms such as anthrax spores that settle on clothing could pose risks if stirred into the air and inhaled.)

As long as one is able to function while wearing special gear, physical protection against chemical or biological agents is realistic, at least for limited periods. Medical protection can be more problematic. Unless administered in advance of or soon after exposure, antidotes to nerve agents and medical responses to biological agents are useless.

23. Charles Piller and Keith R. Yamamoto, "The U.S. Defense Research Program in the 1980s: A Critique," in Wright, *Preventing a Biological Arms Race,* 133.

24. Victor W. Sidel, "Biological Weapons Research and Physicians: Health and Ethical Analysis," *The PSR Quarterly* 1, no. 1 (March 1991), 40.

25. Annual expenditures for overall AIDS research was reported in U.S. Department of Health and Human Services, *Health United States, 1994* (Washington, DC: Government Printing Office, 1995), 239. The ten-year estimate for AIDS vaccine research was extrapolated from figures for the 1990s, provided by Wendy Wertheimer, senior adviser in the AIDS Research Office of the National Institute of Allergy and Infectious Diseases. Personal communication, May 3, 1996.

26. U.S. Department of Defense, *Report on Nonproliferation and Counter-proliferation Activities and Programs,* Office of the Deputy Secretary of Defense, May 1994, introductory letter, ES-2.

27. Gary Taubes, "The Defense Initiative of the 1990s," *Science* 267, no. 5201 (February 24, 1995), 1096. A variety of possible detection techniques were summarized in Richard W. Titball and Graham S. Pearson, "BWC Verification Measures: Technologies for the Identification of Biological Warfare Agents," *Politics and the Life Sciences* 12, no. 2 (August 1993), 255–63.

28. Busbee discussed the biological defense program on July 26, 1995, at a forum of the Chemical and Biological Arms Control Institute in Washington, D.C. Remarks cited here are (with permission) from his presentation and from interviews and communications on July 26 and 27, 1995.

29. Harvey Ko, interview, August 25, 1995.

30. Taubes, 1100.

31. Erhard Geissler and Robert H. Haynes, eds., *Prevention of a Biological and Toxin Arms Race and the Responsibility of Scientists* (Berlin: Akademie-Verlag, 1991).

32. Mark L. Wheelis, "The Role of Epidemiology in Strengthening the Biological Weapons Convention," ibid., 277–81.

33. Mark L. Wheelis, "Strengthening Biological Weapons Control Through Global Epidemiological Surveillance," *Politics and the Life Sciences* 11, no. 2 (August 1992), 181–82.

34. S. J. Lundin, "Global Epidemiological Surveillance: Of General Concern or a Matter for the BWC?" *Politics and the Life Sciences* 11, no. 2 (August 1992), 192; Barbara Hatch Rosenberg, "The Politics of Epidemiological Surveillance," *Politics and the Life Sciences* 11, no. 2 (August 1992), 193; Jack Woodall, "Preparedness Is Nine-Tenths of Prevention," *Politics and the Life Sciences* 11, no. 2 (August 1992), 194.

35. Mark L. Wheelis, "The Global Epidemiological Surveillance System and Vaccines for Peace: Complimentary Initiatives in Public Health and Weapon Control," in *Control of Dual-Threat Agents: The Vaccines for Peace Programme,* ed. Erhard Geissler and John P. Woodall (New York: Oxford University Press, 1994), 176–77.

36. Draft proposal for Global Program to Monitor Emerging Diseases, prepared by ProMED Working Group, Federation of American Scientists, Washington, DC, July 1995, 5.

37. E. Geissler and M. Rice, in Geissler and Haynes, 509. Geissler spoke of internationalizing vaccine development a year earlier in

"The International Control of Biological Weapons," *GeneWatch* 6, no. 1 (1989), 1–4.

38. Ibid., 234.

39. Ibid., 237–38.

40. "Biesenthal Consensus: Biesenthal Vaccine Initiative (Vaccines for Peace)," Biesenthal, September 13, 1992, in *Politics and the Life Sciences* 12, no. 1 (February 1993), 102.

41. Graham S. Pearson, "Vaccines for Peace: An Incentive for Developing Countries," *Politics and the Life Sciences* 12, no. 1 (February 1993), 89.

42. Erhard Geissler, "Vaccines for Peace: A Response to Commentaries," *Politics and the Life Sciences* 12, no. 1 (February 1993), 95.

43. Jack Woodall, "An Exceptional Window of Opportunity: Comments on 'Vaccines for Peace'," *Politics and the Life Sciences* 12, no. 1 (February 1993), 91. Stephen S. Morse, "Vaccines for Public Health: Can Vaccines for Peace Help in the War Against Disease?" in Geissler and Woodall, *Control of Dual-Threat Agents,* 175.

44. Oliver Thränert, "Vaccines for Peace: A Political Scientist's Critique," in Geissler and Woodall, 55. Milton Leitenberg, "The Conversion of Biological Warfare Research and Development Facilities to Peaceful Uses," in Geissler and Woodall, *Control of Dual-Threat Agents,* 105.

45. "Update: Vaccines for Peace," ProCEID—Program for Controlling Emerging Infectious Diseases: Mission Statement, ProCEID Steering Committee, *Politics and the Life Sciences* 14, no. 1 (February 1995), 90–91.

46. Erhard Geissler, personal communication, August 29, 1995.

47. Although the formal mission of the Biological Defense Research Program is to defend military forces, army spokesmen have indicated that it is to protect the civilian population as well. Past biological warfare tests over cities "proved how helpless the United States would be if an enemy were bent on causing problems. Many believe the U.S. is still vulnerable, one reason defensive BW research continues to take place," according to the army's public affairs chief at Fort Detrick. Norman M. Covert, *Cutting Edge: A History of Fort Detrick, Maryland, 1943–1993,* Headquarters U.S. Army Garrison, Public Affairs Office, Fort Detrick, MD, 1993, 58.

CHAPTER 11

1. U.S. Senate Committee on Governmental Affairs, *Global Spread of Chemical and Biological Weapons: Hearings Before the Committee on*

Governmental Affairs and Its Permanent Subcommittee on Investigations,
101st Cong., 1st sess., February 9, 10 and May 2, 17, 1989, 7, 16,
95.

2. John Ellis Van Courtland Moon, "Controlling Chemical and Bio-
logical Weapons Through World War II," in *Encyclopedia of Arms
Control and Disarmament,* vol. 1, ed. Richard Dean Burns (New York:
Scribner's, 1993), 657.

3. Alva Myrdal, *The Game of Disarmament: How the United States and
Russia Run the Arms Race* (New York: Pantheon, 1976), 227–28.

4. William of Malmesbury, *Gesta Regum Anglorum,* translated by Joseph
Stevenson, *William of Malmesbury: A History of the Norman Kings*
(Dyfed, Wales: Llanerch Enterprises, 1991), 84.

5. Moon, 658.

6. Ibid.

7. Robert Ward, *An Enquiry into the Foundation and History of the Law of
Nations in Europe from the Time of the Greeks and Romans to the Age of
Grotius,* vol. 1 (London: Butterworth, 1795), 13, 253.

8. Moon, 659.

9. Stockholm International Peace Research Institute (SIPRI), *The
Problem of Chemical and Biological Warfare: The Rise of CBW and the
Law of War,* vol. 3 (New York: Humanities Press, 1973), 94; Moon,
659.

10. *Protocol for the Prohibition of the Use in War of Asphyxiating, Poisonous
or Other Gases, and of Bacteriological Methods of Warfare.* Signed at
Geneva, June 17, 1925.

11. Moon, 673.

12. J. P. Perry Robinson, "Origins of the Chemical Weapons Convention," in *Shadows and Substance: The Chemical Weapons Convention,*
ed. Benoit Morel and Kyle Olson (Boulder, CO: Westview, 1993),
40.

13. Michael Mandelbaum, *The Nuclear Revolution: International Politics
Before and After Hiroshima* (New York: Cambridge University Press,
1981), 32–33, 37–38.

14. Ibid., 39. Mandelbaum's views are questioned, as is the notion that
the poison taboo was strongly rooted in ancient societies, in
Richard Price, "A Genealogy of the Chemical Weapons Taboo,"
International Organization 49, no. 1 (Winter 1995).

15. U.S. Army Chemical Corps, *Summary of Major Events and Problems,*
U.S. Army Chemical Corps Historical Office, Army Chemical Center, MD, fiscal year 1954, 8–11.

16. *Summary,* fiscal year 1955, 29.

17. *Summary,* fiscal year 1956, 13–14, 40.

18. *Summary,* fiscal year 1960, 6–10.

19. *Summary,* fiscal years 1961–1962, 120.

20. Colonel W. D. Tigertt, "Status on the Medical Research Effort," *Military Medicine* 128 (1963), 142.

21. Elinor Langer, "Chemical and Biological Warfare (I): The Research Program," *Science* 155 (January 13, 1967), 174.

22. Laurie Garrett, *The Coming Plague: Newly Emerging Diseases in a World Out of Balance* (New York: Farrar, Straus and Giroux, 1994), 416.

23. Ibid., 154; John Keegan, *A History of Warfare* (New York: Knopf, 1993), 50.

24. Garrett, 10, 618.

25. Ibid., 486.

26. Stephen S. Morse, "Vaccines for Public Health: Can Vaccines for Peace Help in the War Against Disease?" in *Control of Dual-Threat Agents: The Vaccines for Peace Programme,* ed. Erhard Geissler and John P. Woodall (New York: Oxford University Press, 1994), 168–69.

27. Joseph White, "Health Care Reform the International Way," *Issues in Science and Technology* 12, no. 1 (Fall 1995), 36.

28. The disparity between rich and poor countries is huge. Of the $1.7 trillion in public and private health care expenditures spent throughout the world in 1990, almost 90 percent was in high-income countries. *World Development Report 1993: Investing in Health,* published for the World Bank (New York: Oxford University Press, 1993), 4, 258–59.

29. Konrad Lorenz, *On Aggression* (New York: Harcourt Brace, 1966), 289. See also Robert Hunt Sprinkle, *Profession of Conscience: The Making and Meaning of Life-Sciences Liberalism* (Princeton, NJ: Princeton University Press, 1994), 70–72.

30. Dave Grossman, *On Killing: The Psychological Cost of Learning to Kill in War and Society* (Boston: Little, Brown, 1995), 33–35. To the chagrin of World War I commanders, troops facing each other in nearby trenches were often reluctant to fire, having developed a live-and-let-live system. Robert Axelrod, *The Evolution of Cooperation* (New York: Basic Books, 1984), chap. 4.

31. Grossman, *On Killing,* 250–53. Grossman attributes the Nazi extermination of Jews and others to sadists and thugs who were specially sought for these assignments. Many were identified in the SS ranks through observation of behavior and then assigned to the death

camps. Two or three percent of men in any population evidently have no compunctions about killing. Grossman, *On Killing,* 78–79, 180–181. See also *Diagnostic and Statistical Manual of Mental Disorders,* 4th ed. (*DSM-IV*) (Washington, DC: American Psychiatric Association, 1994), 648–50. These issues remain controversial. In the well-known Milgram experiments, most subjects acceded when instructed to administer (perceived) painful electric shocks to others as part of a presumably legitimate experiment. When the overseeing authority was not physically present, however, subjects commonly disobeyed the instruction. Stanley Milgram, *Obedience to Authority* (New York: Harper Colophon, 1975), 59–62.

32. Grossman, *On Killing,* 39.

33. Ibid., 332.

34. Daniel Patrick Moynihan, "Defining Deviancy Down," *The American Scholar* (Winter 1993), 19, 30.

35. Leon R. Kass, "Organs for Sale? Propriety, Property, and the Price of Progress," *The Public Interest,* no. 10 (Spring 1992), 68–70.

"Which would worry you more," I asked a friend on a whim in 1995, "being attacked with a biological weapon or a chemical weapon?" He looked quizzical: "Frankly, I'm afraid of Alzheimer's." We laughed. He had elegantly dismissed my question as an irrelevancy. In civilized society people don't think about such things.

The next day, the nerve agent sarin was released in the Tokyo subway system—in Japan, no less, one of the safest countries in the world. I called my friend and we lingered over the coincidental timing of my question. A seemingly frivolous speculation one day, a deadly serious matter the next.

Some analysts, unsurprised by the Tokyo attack, expect chemical and biological weapons to be used more often in the future. Others are less convinced. They see no inherent reason that these weapons be used, especially if treaties and other disincentives are implemented.

In assessing these and other questions related to biological and chemical warfare, I have spoken with hundreds of people— scientists, army personnel, politicians, government and non-government experts, people who have suffered because of army testing. Some were harshly critical of U.S. policies, and others enthusiastically supportive. While not concealing my views, I have sought to project the full range of thinking on the issues.

Nevertheless, one cannot avoid compassion for individuals who were made ill by biological and chemical warfare testing— or for the families of victims who were killed by them. In some cases injuries were presumptive, in others irrefutable. For sharing their experiences and concerns with me I am grateful to Elizabeth Barrett, Earl Davenport, Diane Gorney, Richard Meixner, Edward J. Nevin III (and members of his family), and

Carol Thomas. I thank Neil Levitt and Helen Ramsburg, former employees in the biological defense program, for their perspectives on the program.

I also benefited from discussions about a variety of biological and chemical warfare issues with several army officials and employees, including Colonel Arthur Anderson, former Brigadier General Walter Busbee, Colonel David Franz, former Colonel David Huxsoll, Carol Linden, Cheryl Parrott, Major Dale Vander Haam, and Brigadier General Russ Zajtchuk. Despite some disagreements, conversations with these officials were invariably cordial.

In the area of biological and chemical arms control, I spoke with many who are close to the issue. Among them are Richard D'Andrea, Amy Gordon, Martin Lancaster, Donald Mahley, and William Staples, all currently or recently with the U.S. Arms Control and Disarmament Agency; Barbara Rosenberg of the Federation of American Scientists; and Kathleen Bailey and Frank Gaffney, former U.S. government officials. Valuable insights came as well from Graham Pearson and John Walker of Britain's Chemical and Biological Defense Establishment, Amos Radian of the Israeli Foreign Ministry, and Yehiel Yativ, an ambassador at the Israeli mission to the United Nations. Timur Alasaniya, UN political affairs officer, and Sohrab Kheradi, deputy director of the UN Center for Disarmament Affairs, facilitated my attending a 1994 Geneva conference on biological arms control.

Several members of Congress and their staffs have expressed particular concern about biological warfare issues, especially U.S. testing. On this subject, I have talked with, or testified at hearings conducted by, former Representative Wayne Owens, Representatives Martin Sabo and John Conyers, and Senators John Glenn, John D. Rockefeller IV, and Paul Wellstone. James Tuite, an aide to retired Senator Donald Riegle, has provided helpful perspectives about Gulf War syndrome, as have Diana Zuckerman and Patricia Olson, who worked for Senator Rockefeller.

Senator Frank Lautenberg offered strong support for an advertisement that I organized urging the U.S. Senate to ratify the Chemical Weapons Convention (*New York Times*, November 4, 1995). For helping to identify and contact the 64 prominent scientists, diplomats, business figures, and retired army generals who signed (and paid for) the ad, I thank Anne H. Cahn of American University and Victor Rabinowitch of the MacArthur Foundation.

I have enjoyed conversations with many scientists and physicians whose sensitivity to ethics and biological warfare issues has helped inform mine. They include Kenneth Buchi, Nancy Connell, Naomi Franklin, Erhard Geissler, Jay Jacobson, Jonathan King, Zell McGee, Stephen Morse, Meryl Nass, Barbara Rosenberg, Victor Sidel, Robert Sprinkle, David Thaler, and Jack Woodall.

Several experts read all or portions of the book in draft form and made many valuable suggestions: Seth Carus, Nancy Connell, Marie Isabelle Chevrier, Matthew Meselson, John Ellis van Courtland Moon, and Raymond Zilinskas. Comprehensive comments by Julian Perry Robinson and Jonathan Tucker were particularly helpful in steering the text toward clarity and accuracy.

I thank the staff at W. H. Freeman for their energetic support, especially Elizabeth Knoll, as sensitive and talented an editor as one could wish for. Her ideas have enriched this book throughout. And I am grateful to the Rockefeller Foundation for inviting me to finish the book as a scholar-in-residence at its enchanting Villa Serbelloni in Bellagio, Italy.

Finally, I thank my wife, Ruth, for the many good things in my life, and for her suggestion that the book be titled *The Eleventh Plague*.

Labor Department, U.S., 64
Lancaster, Martin, 184–185
Leavitt, Michael O., 73
Leighton, Philip A., 24–27
Leishmaniasis, 129–130
Levin, Carl, 45–46, 49
Levitt, Neil, 47–52, 56
Lewis, Flora, 93
Libya, 5, 83–84, 193
Lieberman, Joseph, 94, 213
Linden, Carol, 51, 55
Livingstone, Neil C., 12, 165
Lockheed Corporation, 67
Long, David, 163–166
Lorenz, Konrad, 220–221
Lysergic acid, 18, 31–33

MacArthur, John R., 80
McGee, Zell, 66, 72
McGovern, George, 175
Mahley, Donald, 128, 173–175, 191
Malkiam, Yitzhak, 112
Mamats, 105, 115–116
Mandelbaum, Michael, 216–217
Manu Law, 214
Marijuana, 31
Martin, Edward, 136–137
Masahata, Akio, 2
Mathias, Charles, 51
Matsumoto gas attack, 151–152,
    154–155
Media, growth, 143–144, 146, 155,
    159–160
Medical Research Institute of
    Infectious Diseases, 40, 47–57,
    97
Mehta, Zubin, 114
Melling, Nancy, 73
Mescaline, 17, 31–32
Meselson, Matthew, 13
Meyer, Bernd, 82
Middle East Watch, 92
Mikulak, Robert, 192
Miller, Barry, 30
Minneapolis, Minnesota, 18, 25,
    28, 160
Mirzayanov, Vil, 193

Moodie, Michael, 151, 192
Moon, John, 215–216
Moore, Molly, 99–101
Moral aspects, 8–9, 37–41, 94, 129,
    136, 138, 161, 165–166, 171,
    194–195, 213–218, 221–225
Morris, Harvey, 89–90
Morse, Stephen, 220
Mosquitoes, 28–31
Moss, Frank, 72
Mottice, Susan, 74
Moynihan, Daniel Patrick, 222
Murphy, Frances, 130–131
Murrain, 227n1
Mustard gas, 2, 82, 88–89, 126, 142,
    158, 187
Myasthenia gravis, 136
Mycotoxin T-2, 71
Mycotoxins, 179

Naidus, Edward, 158–159
National Cancer Institute, 54
National Disaster Medical
    System, 167
National Institute of Medicine, 138
National Institutes of Health, 76,
    203
Nazis, 37, 113, 271n31
Nerve gas, 2, 7, 12, 60, 62, 70, 82,
    88–89, 106, 111, 113, 115, 126,
    133–135, 139, 141–142,
    151–155, 158–159, 161–162,
    169, 177, 221, 283
Netherlands, 81–82
Neurons, 135
Neurotransmitters, 135
Nevin, Edward, 17
New York City, 18, 24, 160, 167,
    170
New York State Psychiatric
    Institute, 31
Nixon, Richard, 9, 12, 176–177,
    199
North Atlantic Treaty
    Organization, 12, 211
North Korea, 5, 80, 153, 242n7
North Vietnam, 131